Scientific Illustration

A Guide to Biological, Zoological, and Medical Rendering Techniques, Design, Printing, and Display

Phyllis Wood

VNR
VAN NOSTRAND REINHOLD COMPANY
New York Cincinnati Toronto London Melbourne

To all my students, past, present, and future. *Learning to draw is learning to see.*

First published in paperback in 1982

Library of Congress Catalog Card Number 78-2834
ISBN 0-442-29307-0

Printed in the United States of America
Designed by Loudan Enterprises

Van Nostrand Reinhold Company Inc.
135 West 50th Street, New York, NY 10020

Van Nostrand Reinhold Australia Pty. Ltd.
480 Latrobe Street, Melbourne, Victoria 3000

Van Nostrand Reinhold Company Ltd.
Molly Millars Lane, Wokingham, Berkshire, England
RG11 2PY

Cloth edition published 1979 by Van Nostrand Reinhold Company Inc.

16 15 14 13 12 11 10 9 8 7 6 5 4 3 2

Contents

Preface 5
Chapter 1. Perspective 6
Chapter 2. Drawing 12
Chapter 3. Light and Shadow 26
Chapter 4. Black-and-white Drawing 35
Chapter 5. Continuous Tone 51
Chapter 6. Color 67
Chapter 7. Animal Illustration 78
Chapter 8. Diagrams 84
Chapter 9. Design and Layout
Chapter 10. Printing for Publication 116
Chapter 11. Exhibits 123
Chapter 12. Career Guide 142
Bibliography 146
Index 147

Acknowledgments

I first started working on this book in an independent and solitary manner, depending entirely on my own resources for the information and illustrations. As the book developed, however, contributions, advice, and support from my colleagues and friends became more and more important, resulting in an immense enrichment of these pages.

My first acknowledgment is to my students, not only those whose illustrations I have included but all those whom I had the privilege of guiding through a part of their artistic development.

I also wish to thank Jessie Phillips Pearson, first Director of the Department of Health Sciences Illustration at the University of Washington, who was an inspiration to work with for many years. Kathleen Schmitt has been my steadfast editor, guiding me through the paths of participles. Lee Haines and Michael McIntosh are responsible for the original photography and color transparencies not otherwise credited.

Two organizations to which I am privileged to belong are the Association of Medical Illustrators and the Guild of Natural Science Illustrators. Their standards of excellence and generous sharing of information have been of great value in the development of this book.

The experts who graciously reviewed portions of the text contributed their many talents and cumulative knowledge. Grover Gilbert, architect, dealt with the areas on perspective. Joel Ito, Director of Medical Illustration at the Oregon Regional Primate Research Center, shared his expertise with the airbrush. Janet MacKenzie, scientific illustrator, reviewed coquille-board technique. Cheryl Vigna, designer, added her special brand of creativity to the chapter on design and layout. Bert Hagg, Director of Printing at the University of Washington, reviewed the technicalities of printing and production methods. Professor Thomas A. Stebbins, Director of Health Sciences Illustration, made many thoughtful contributions to the chapter on exhibits. Russell W. Newman, attorney, helped to clarify the critical areas of career guidance and contracts. The Regional Primate Research Center and its Assistant Director, Douglas M. Bowden, M.D., and the Health Sciences Learning Resources Center and its Director, Robert S. Hillman, M.D., gave me both direction and support.

Preface

Scientific illustration is produced for a specific kind of visual communication in the sciences. This communication can pass from scientist to colleague, teacher to student, or research foundation to layman. The artist must therefore be aware of the viewer's level of knowledge and must relate the message in a logical sequence without confusing him with too much or too little information. The art must be rendered with scientific accuracy and artistic integrity within the production framework of its ultimate use: print, projection, or display.

This book is directed to the artist/scientist who wishes to produce bioart for print (books and publications), projection (slides, television, motion pictures), or display. The emphasis is not on the original piece of art but on the final form in which it will be used. In order to control the production process, the artist must understand the mechanics of transferring the original art to paper or transparency form. He must be able to communicate clearly with the typographer, the platemaker, the printer, and the photographer, using the correct technical terminology. He must understand the capabilities and limitations of these technical processes and be able to prepare his artwork to take advantage of them. He must know why most artwork is reduced rather than enlarged, how much information can be put on a slide or a printed page, and how to handle overlays and screens.

Although scientific illustration is serious and factual, it need not be handled in a dry or monotonous style. It can and should be designed as thoughtfully and innovatively as any piece of art, using current styles and methods and combining traditional with advertising and computer techniques. At present there is a tremendous proliferation of scientific Information, both written and spoken, that is competing for an audience. This means that the illustrator has the important responsibility of creating a greater number of accompanying visuals that will attract viewers, communicate clearly and quickly, and be easy to remember. This book is designed to enable him to meet these needs.

CHAPTER I.

Perspective

1-1. Theban tomb painting. *Guests at a Feast.* C. 1411–1375 B.C.

Perspective drawing is learned, not instinctive. Unless trained in the rules of perspective, an artist will draw what he knows, not what he sees. Ancient Egyptian drawings are beautiful examples of this tendency (1-1). The body is drawn in the most honestly recognizable position: the eye viewed from a frontal plane, the nose in side view, the upper torso from the front, the lower torso from the side, the hands from the back, and the feet in profile. There are few overlapping figures or objects, and one object does not recede behind another. The size of a person or thing does not relate to its proximity to the viewer but to its importance. Kings were drawn very large; slaves and women were drawn small.

Many of these same characteristics appear in primitive paintings of the 19th and 20th centuries (1-2) and in children's drawings (1-3). Modern art may deliberately defy the rules of perspective in search of another truth (1-4). Plato wrote that perspective was "a kind of trick which took advantage of the weakness of our senses, for in reality the width and length of a bed or table do not contract, but remain constant, as mathematical measurements show. . . . Painting, therefore, is busy about a work which is far removed from the truth."

DEFINITION

Perspective may be defined as the appearance of reality and is determined by the position from which it is observed. Every change in the observer's viewpoint changes all the perspective relations. The eye can see the object in its true shape only when it is parallel to the face or picture plane. In any other position the object is foreshortened (1-5). The elements of linear and atmospheric perspective that govern the drawing of all objects are: relative size, relation of angles, overlapping, and distinctness. These elements are examined in the exercise presented later in the chapter.

1-2. Flora Fryer. *Drummer Boys.* Acrylic, 32" × 22", 1968.

1-3.

1-4. Jacob Lawrence. *The Swearing In.* Serigraph, 1977. Executed at the request of President Jimmy Carter.

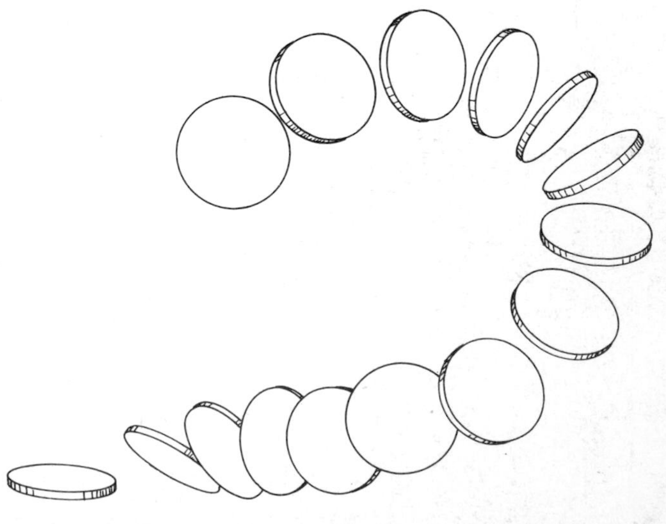

1-5.

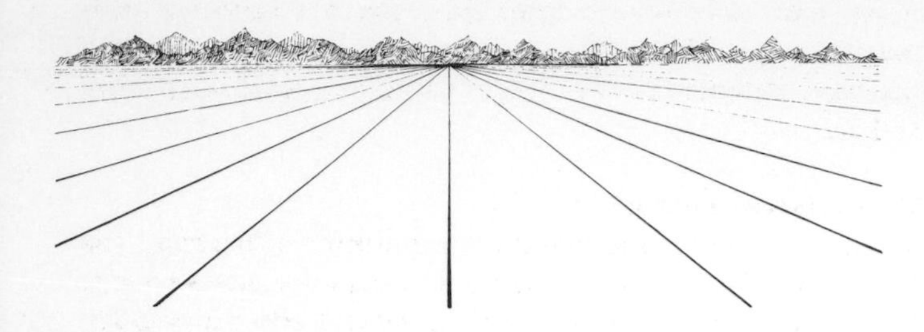

1-6. Reed Eastman.

1-7. Phyllis Wood.

RULES

The rules governing perspective that concern the scientific illustrator are simple.

1. Receding parallel lines seem to converge away from the eye to a common vanishing point (1-6).
2. Objects appear smaller in relative proportion to their distance from the eye (1-7).
3. Surfaces that are parallel to the picture plane appear in their true shape (1-8).
4. Surfaces are foreshortened in relative proportion to their angle away from the picture plane (1-8).
5. A circle that is parallel to the picture plane appears as a circle (1-9).
6. A circle observed at an angle to the picture plane is foreshortened and appears as an ellipse; in the most extreme foreshortening the circle appears as a line (1-9).
7. An object appears less distinct in proportion to its distance from the eye (1-10).

THE PICTURE PLANE

The picture plane is an imaginary plane between the viewer and the object. Imagine a pane of glass hovering over the object at an angle parallel to the face and at right angles to the line of sight. It is on this plane that the object is measured (1-11). If the object were traced onto the surface of the glass, it would be drawn in correct linear perspective.

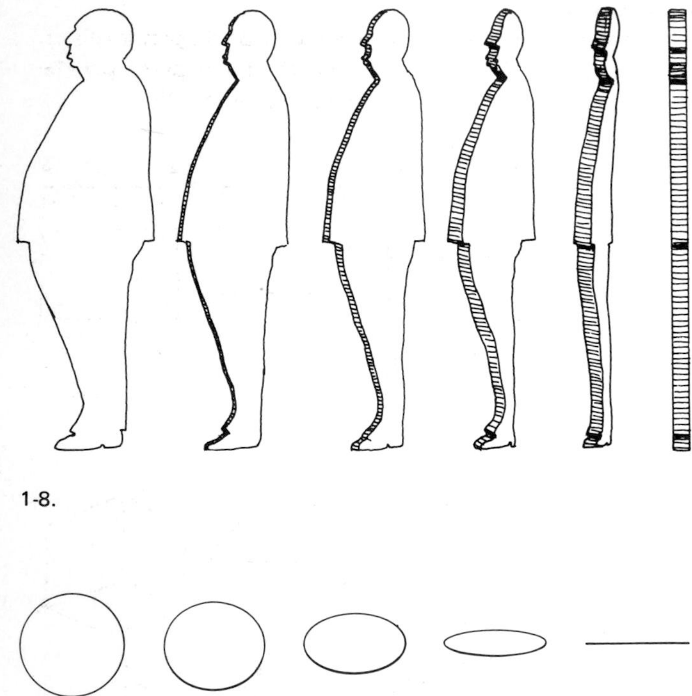

1-8.

1-9.

1-10.

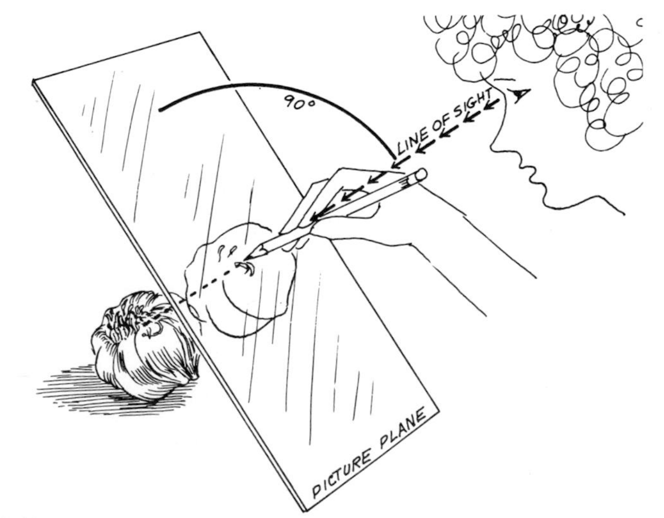

1-11.

EXERCISES

We are going to explore perspective by using the basic shapes of the cube and the cylinder. These shapes illustrate the elements of perspective that are relevant to the scientific illustrator and can be related to biologic shapes. With complicated shapes you can fool yourself into believing that your drawing is accurate, but any error in a simple geometric shape is immediately apparent. If we first learn to observe and to draw solid geometric shapes that are stable, precise, and predictable, we can transfer that ability to biologic shapes that are variations and combinations of geometric shapes.

In drawing small subjects such as these the perspective variations in foreshortening and angle change are very slight. Careful and accurate interpretation of these subtleties of measurement is the purpose of this precise way of drawing. Such subtleties, drawn without exaggeration, produce the realistic look that is the goal of the scientific illustrator.

The cube

The 2″ cube is a good subject with which to start studying perspective. If you draw it accurately, your illustration can be turned sideways or upside down and still appear as a realistic equal-sided cube.

One-point perspective

Obtain a cube that measures about 2″ on each side. Draw it with one side parallel to the picture plane. Since this front side is parallel to the picture plane, it will be seen in its true shape, as a square. None of the other sides is visible (1-12).

Pretend that the cube is transparent and draw the other sides. The back side is smaller because it is further away. The other four sides are foreshortened and are bordered with parallel edges. These four parallel edges converge toward a common vanishing point. This is called one-point perspective (1-13).

Two-point perspective

Now turn the cube so that you can see some of the top. The front side is no longer exactly parallel with the picture plane. It is slightly foreshortened, and its vertical edges eventually converge. Parallel lines converge away from the eye: the vertical edges parallel to each other in this drawing therefore converge downward to a vanishing point. The lateral edges on the top of the cube are also parallel to each other and converge upward away from the eye toward a second vanishing point (1-14).

If the cube were transparent, you could see that all parallel edges converge toward these two respective vanishing points. The plane that is furthest from the eye (at top or bottom) is proportionately wider. This is because it is at a smaller angle to the picture plane (1-15).

As you move the cube, viewing more and more of the top, the angles and surfaces change in relation to their distance from and angle to the picture plane. The only constant is the closest or leading edge, because it is the only part of the cube that is on the picture plane.

Three-point perspective

Turn the cube so that only one corner is on the picture plane. Every surface is foreshortened, and three groups of parallel lines converge towards three vanishing points (1-16).

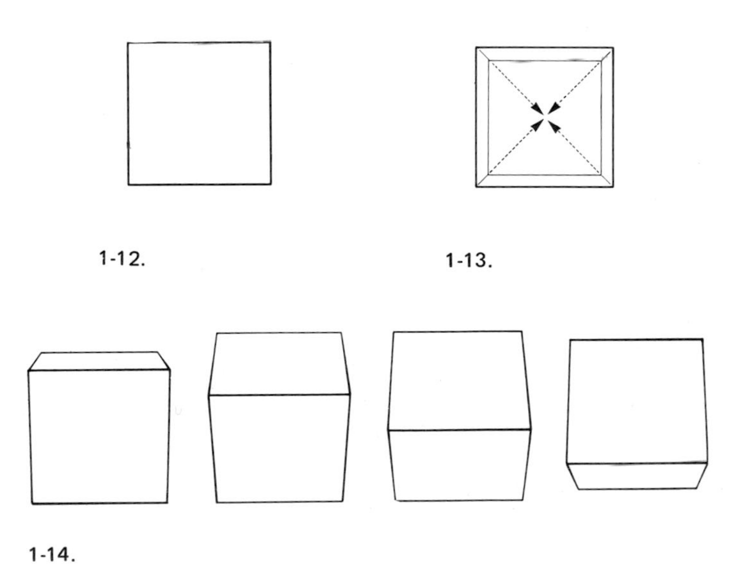

1-12. 1-13.

1-14.

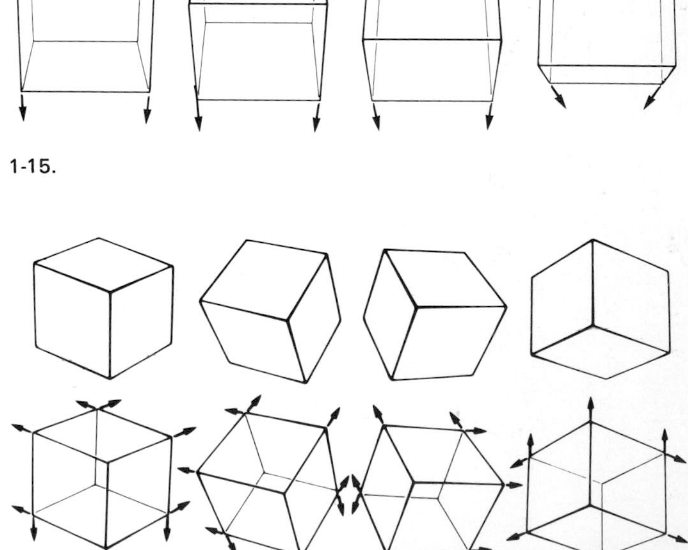

1-15.

1-16.

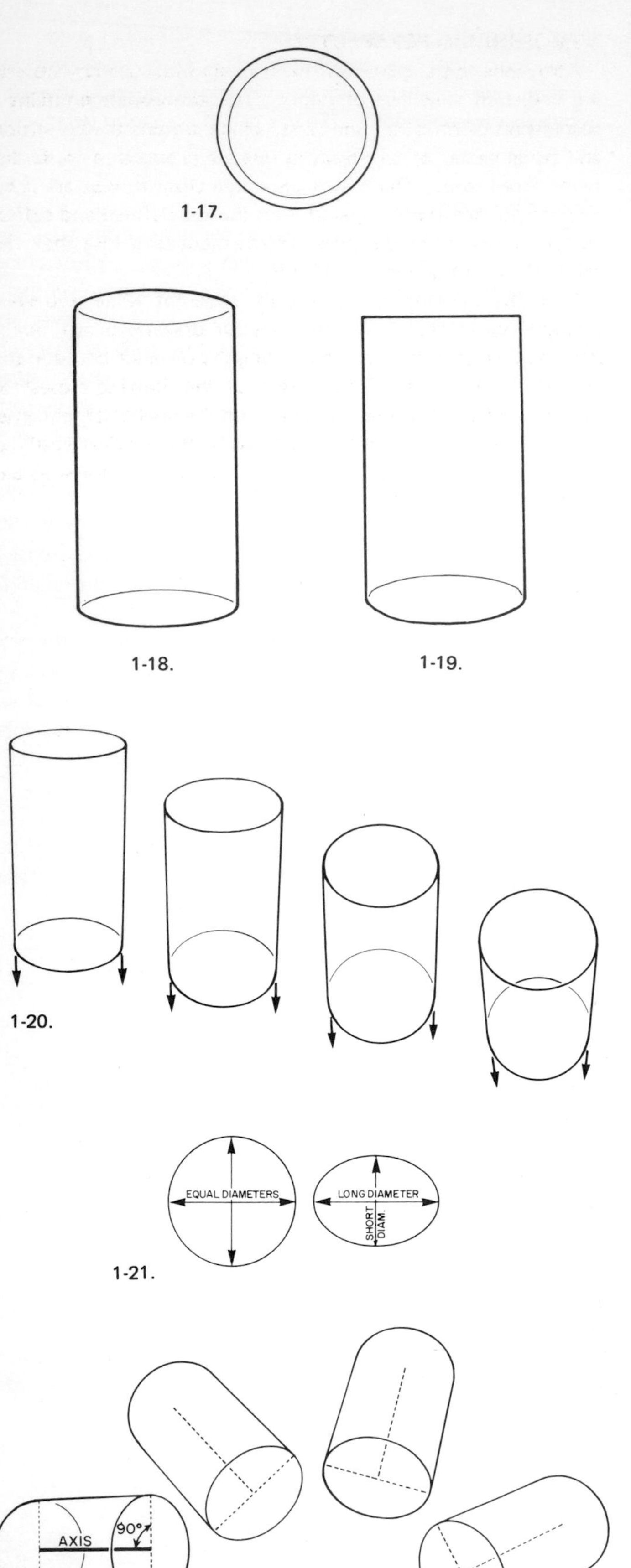

1-17.

1-18.

1-19.

1-20.

1-21.

1-22.

The cylinder

The cylinder, which includes the elusive ellipse, is often translated into biological shapes. You should achieve an intimate familiarity with this shape so that you do not need to think consciously about the rules that govern the perspective views as you draw it.

Top view

Obtain a cylinder that is easily held in the hand. View it from the top. It appears as a circle, the only part on the picture plane. Imagine that it is transparent. The further rim is a smaller circle because it is further from the eye. Neither one is foreshortened because both are parallel to the picture plane (1-17).

Side view

Observe the cylinder from the side with your line of sight centering on the middle of the long axis. You cannot see either the top or the bottom of the cylinder. Both top and bottom edges are curved, representing the front parts of ellipses. Draw the cylinder as if it were transparent (1-18). The ellipses are very shallow. Lower the cylinder until the top edge centers on the line of sight. It appears as a straight line, while the lower edge is rounder because it is further from the eye and therefore at a smaller angle to the picture plane (1-19), illustrating the rule concerning parallel ellipses that the one further from the eye appears rounder. Drop the cylinder gradually and study the increasing roundness of the ellipses. Raise the cylinder: you will notice the same result. The vertical sides of the cylinder recede from the eye and therefore seem to converge away from the eye (1-20).

Oblique view

Place the cylinder at an oblique angle and observe the same changes (1-22). The cylinder does not change shape if it is observed at an oblique angle. The same drawing, if done accurately, can be turned in any direction and still be correct.

The ellipse

Always remember that an ellipse is a foreshortened circle. Circles do not have corners: the peripheral ends of the ellipse must therefore be rounded because it is evolved from the circle shape.

The circle can be divided into four equal parts with two equal diameters. The ellipse can also be divided into four equal parts but with a shorter and a longer diameter (1-21). A good way to check the accuracy of an ellipse is to draw the short and long diameters and to see whether each quarter is identical in size and curvature.

An ellipse that is part of a cylinder has a stable relationship with the cylinder. The axis of the cylinder is represented by a measurement of its exact center. This axis is always at a right angle to the long diameter of the ellipse (1-22).

In drawing only part of an ellipse it is helpful to check your accuracy by completing the figure and by drawing the axis of the cylinder and the long and short diameters of the ellipse.

1-23. Johsel Namkung.

ATMOSPHERIC PERSPECTIVE

Atmospheric perspective is most easily illustrated by observing a distant view out of doors. The atmosphere contains a suspension of moisture and dust, which softens the definition and tonal values of an object in relative proportion to its distance from you. The leaves on a tree close to you are crisp and deeply colored compared with the less defined and softer-valued leaves in the distance; distant mountains lose their detail and are soft and grayed (1-23).

This phenomenon is not readily apparent when you view a single object that is resting on your drawing board, but it is a valuable principle in interpreting the distance between the eye and a specimen. Those parts of the drawing closest to your eye have the deepest shadows and the brightest highlights. They are the crispest and are defined by the heaviest outlines. More distant structures are less defined, and the tonal values are closer (1-24).

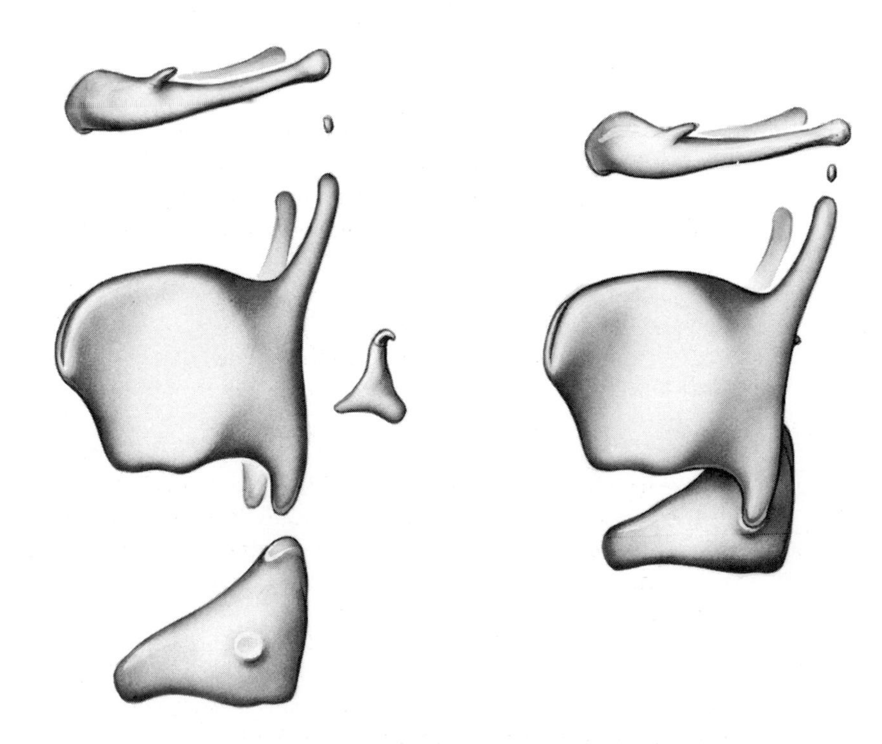

1-24. Lateral view of the hyaline cartilages of the larynx, airbrush, Robert J. Demarest. From *Laryngeal Biomechanics,* B. Raymond Fink, and Robert J. Demarest, Harvard University Press, © 1978 by the President and Fellows of Harvard College.

CHAPTER 2.

Drawing

Inherent in the word "scientific" is the word "accurate": scientific illustration is by definition accurate drawing. The goal of the scientific illustrator is to show the observer the same image that he had when he looked at the specimen. This is not merely to say that the observer should get an impression of the subject: he would get that if he were looking at a photograph, sketch, or abstract interpretation of the subject. The observer must be informed so completely and precisely that, when he looks at the drawing, he is as aware of and enlightened about the subject as if he had seen it himself. While this kind of drawing requires disciplined precision, it must also be artistically pleasing. Meticulous accuracy and aesthetics should be combined.

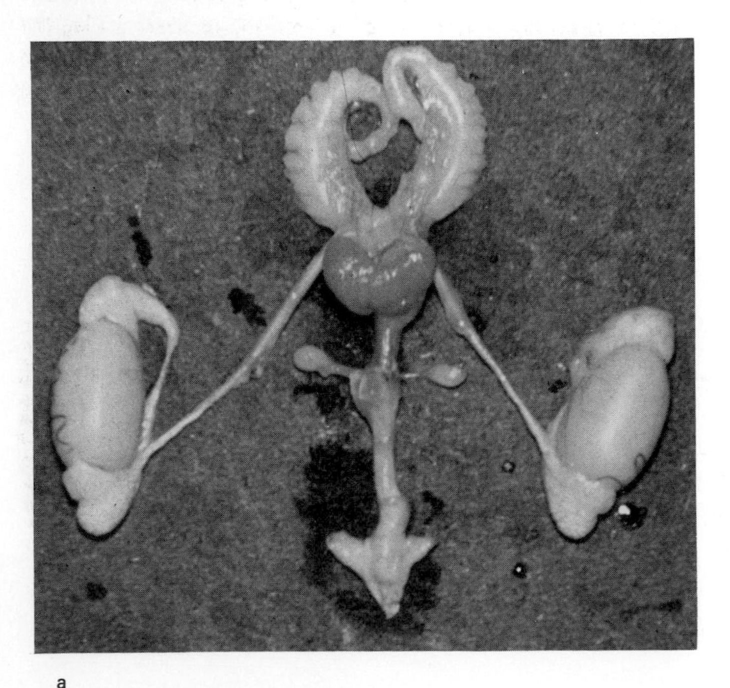

a

Kidney
Ureter
Colon
Seminal vesicle (reflected back)
Coagulating gland
Dorsal prostate
Vas deferens
Ventral prostate
Cowper's gland
Head of epididymis
Testis
Tail of epididymis
Urinary bladder
Urethra
Preputial gland
Penis
Anus
P. Wood
RAT

b

2-1. Photograph and drawing of same subject, male rat urogenitals. a. Miriam Barnes, dissection. b. From Glover Barnes. "Antigenic nature of male accessory glands." *Biology of Reproduction,* 1972, vol. 6, p. 385. Wash on cold-press illustration board.

DIFFERENCES BETWEEN ILLUSTRATION AND PHOTOGRAPHY

Modern scientific photography can document and record a subject accurately and beautifully, but it is the job of the illustrator to interpret it. The artist omits extraneous detail, he clarifies, and he selects. He can dramatize or emphasize the important parts, taking care not to exaggerate or distort for the sake of emphasis or design. The artist can simplify or summarize the essence of a subject and can ghost in what is inside or underneath it. He can reconstruct shards and pieces into a unified whole. He can idealize, ignoring specimen differences. He can recreate the vitality of a living specimen from a dead one. While the camera establishes the existence of a subject, the illustrator illuminates its essence (2-1).

2-2.

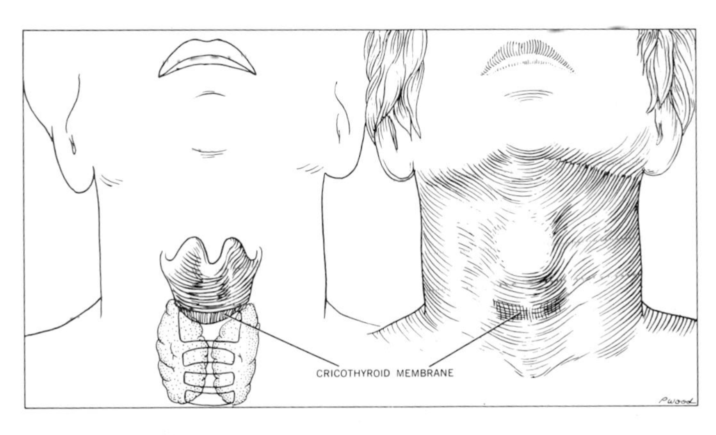

2-3. From Cynthia J. Leitch and Richard V. Tinker, eds. *Primary Care.* F. A. Davis Co., 1978. Ink on paper.

COMMUNICATING WITH THE VIEWER

In communicating with the viewer the artist assumes that they share certain basic perceptions. He must assume that the observer knows that an outline often separates positive and negative space (2-2). Of course, there is no such thing as a line around a specimen. Turn a sphere around and try to fix your eye on the "outline." The outline is one of the many conventions with which we are familiar without even being aware of them. Shading with lines or dots and the principles of perspective are others. A person who is not used to recognizing these codes or devices, however, would not be able to relate them to the subject. To such a person a drawing might be just a pile of squiggles, while to someone else it would represent a meaningful three-dimensional subject.

Another aspect of the understanding between the artist and the observer is that it is not necessary to draw everything. If only the head and shoulders of a person are drawn, we do not assume that they have been amputated from the rest of the body. The observer mentally adds a generalized body to the head and neck. In the same way the observer, if given enough clues, completes any drawing in his mind (2-3).

The scientific illustrator draws only enough to give a true and complete picture of the subject. What is included in a true and complete picture varies greatly, depending on the audience. A less informed audience generally requires more information in an illustration than does a more informed one. Plates reproduced in color or modeled in continuous tone that include many familiar landmarks are commonly used to teach beginning students. The same subject can be illustrated by a simple outline drawing, a portion of a section, or even a diagram when dealing with a well-informed and sophisticated audience (2-4). The artist simplifies and summarizes a subject according to the capacity of the audience to understand it.

a

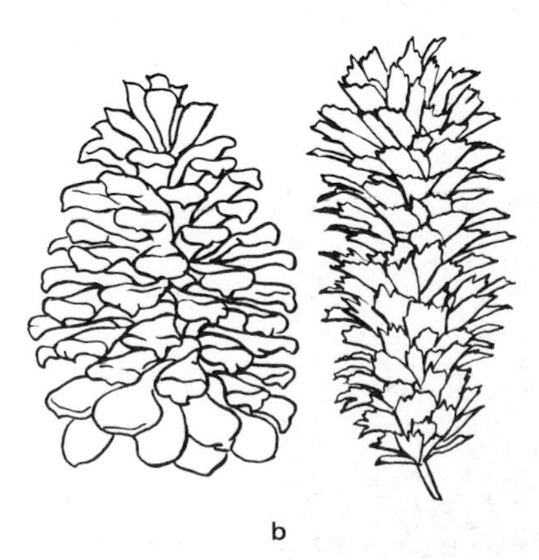

b

c

2-4. a. Reed Eastman. Graphite pencil on paper. b. Phyllis Wood. Pen-and-ink. c. Phyllis Wood. Ink line with shading film.

THE STUDIO

Any serious artist should set aside a special room or part of a room in which he can control his materials and surroundings to the best advantage. It is difficult enough to draw without having to fight the environment.

Your chair should be positioned at a comfortable height and should have a back. Your drawing board should tilt slightly and be set at a height that feels right to you. It should have a smooth, hard surface and be clean and uncluttered. Of course, we all delight in a window that provides filtered north light. This is not always possible. Take advantage of all the natural light you can get plus some good general room light and a single well-directed desk lamp. The primary direction of the light should be from the left (or from the right for left-handed people) so that the shadow of the hand does not fall on the drawing area. A flat cabinet or table within easy reach can hold the necessary tools (2-5).

For basic drawing you will need some medium-hard pencils. If the pencil is too hard, the line will be gray; if it is too soft, the line will not be precise. An HB or #3 is preferable. Paper should be smooth for crispness of line. It should be rather thin so that a drawing can be revised in stages on several overlays and still retain the good parts of the first stages. Use a Pink Pearl or art-gum eraser.

Wash your hands thoroughly with soap and water, put some soft music on the radio, refrain from drinking coffee or other nerve poisons, and start to work.

2-5.

THE SUBJECT

Pick the subject up in your hands; turn it around, *look* at it, and *feel* its shape, its texture, the rhythms of its patterns; notice the variations in shape and size among its primary and secondary parts. All biological material is made up of interrelated parts (2-6), and growth patterns give you many hints about their relationships. Become familiar with all sides of the subject. Although you draw only the near side, you must know what is on the other side, inside, and underneath in order to interpret the subject accurately. This is one of the reasons why drawings done exclusively from photographs are not successful. Photographs can be a valuable aid in drawing, but, when used as the only reference by an artist not familiar with the subject, the results will not stand up. The artist's ignorance will be obvious.

Analysis

Squint at your subject. This may seem to be a strange thing to do, because you see it less clearly. You can, however, see the abstracted essence of the specimen—it is broken into simple areas of dark and light through the veil of the eyelashes. Basic patterns emerge. The main form appears more clearly, with the details screened out (2-7a). Squint at your subject often during all stages of the drawing. This counteracts the natural tendency to become enraptured with detail at the cost of the integration of the subject as a whole (2-7b).

Rhythm and pattern

In order to interpret what you are looking at, try to identify the patterns of growth: the spirals and successions of curves, the methods of branching, the unequivocal design of the mosaics, the rhythms and repetitions, the symmetries and asymmetries. There is always some kind of inherent order in shapes, textures, and colors. Nothing in nature is random, although it may seem so at first glance. Conversely, none of the rhythms or patterns is mathematically precise: no line is perfectly straight, no progression is logarithmically perfect, no curve is geometrically exact. To alter them so that they are perfect would destroy the authenticity of your drawing, making it look artificial and mechanical (2-8).

The pine cone has two oblique spirals, a vertical guide, and another group of circumference guides (2-9). The seemingly random venation of these leaves is apparently the most economical method of feeding its cells (2-10). Symmetric branching contrasts with asymmetric branching. The giant tortoise is a combination of various mosaics arranged in a consistently predictable pattern. Every tortoise has a similar pattern, but each individual tortoise and each individual mosaic are different (2-11). The murex shell has many intricate interrelationships, repetitions, and variations (2-12). Mirror-image symmetry combines with asymmetry of surface pattern in the metallic wood-boring beetle (2-13). All these patterns must be viewed in terms of structure and function. Even apparently decorative parts have some underlying explanation (2-14).

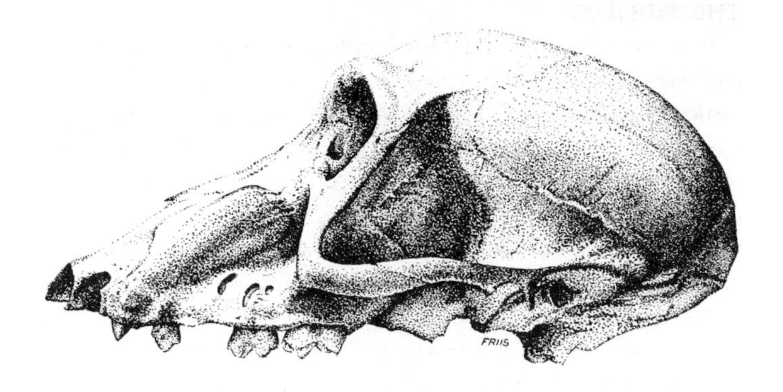

2-6. Anton Friis. Skull. Ink on plate-finish paper.

2-7a.

2-7b. Jayne Lilienfeld. Indian grease bowl. Carbon dust on film.

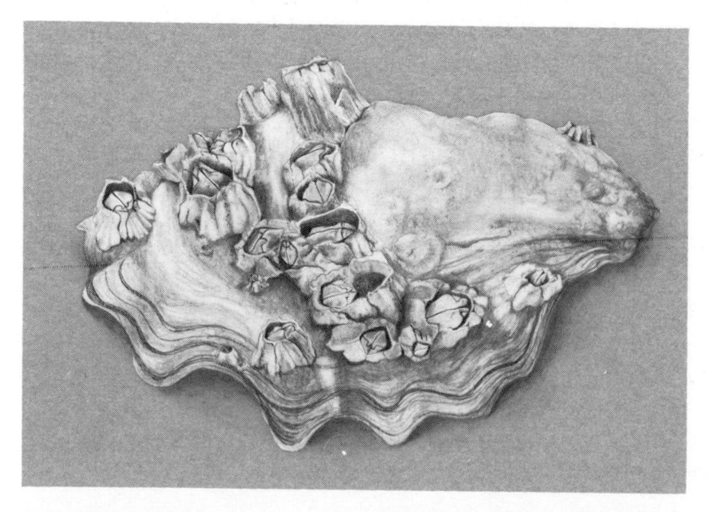

2-8. Nancy Williams. Oyster shell with barnacle. Pencil, watercolor, and ink on frosted film.

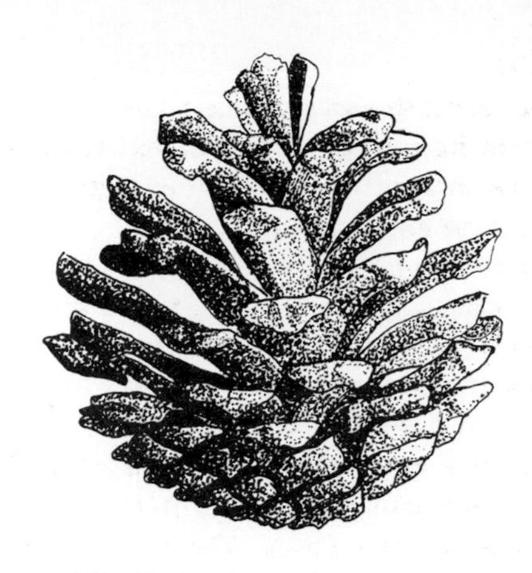

2-9. Phyllis Wood. Ink on paper.

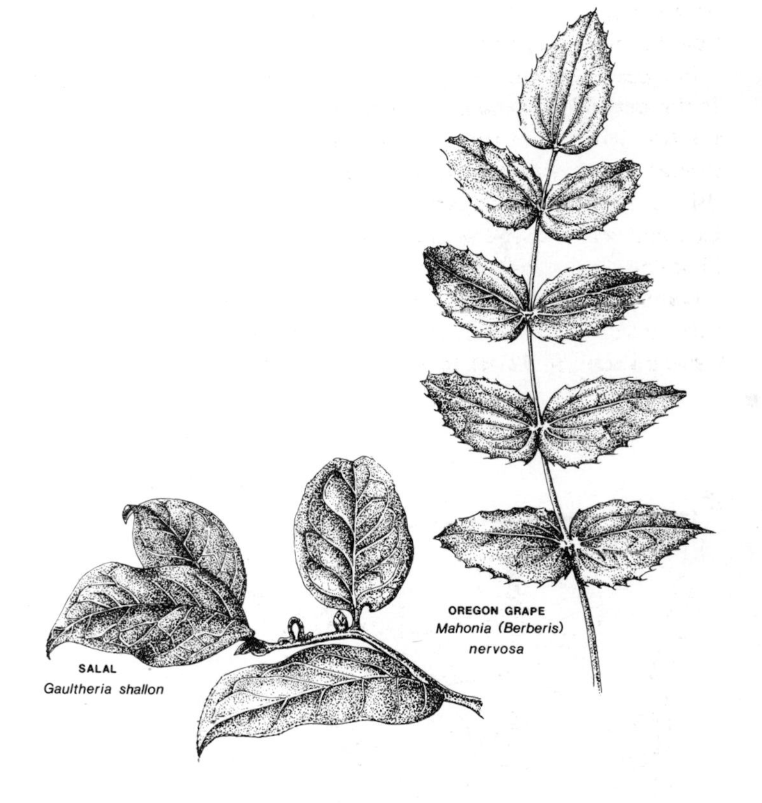

2-10. Laura Dassow. Ink on paper.

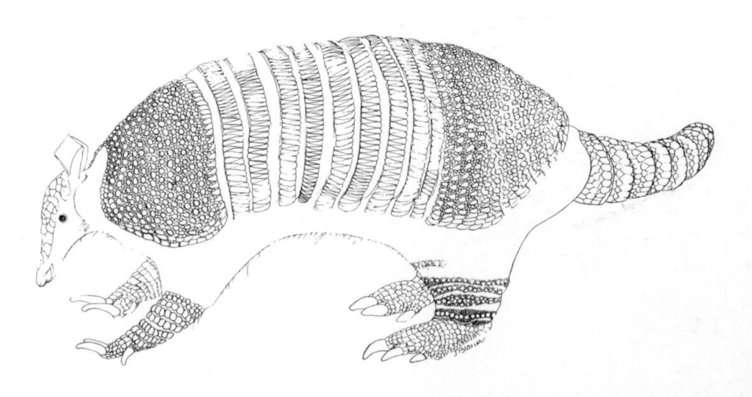

2-11. Janis Blauer. Ink on paper.

2-12. Laura Dassow. ***Murex***. Ink on paper.

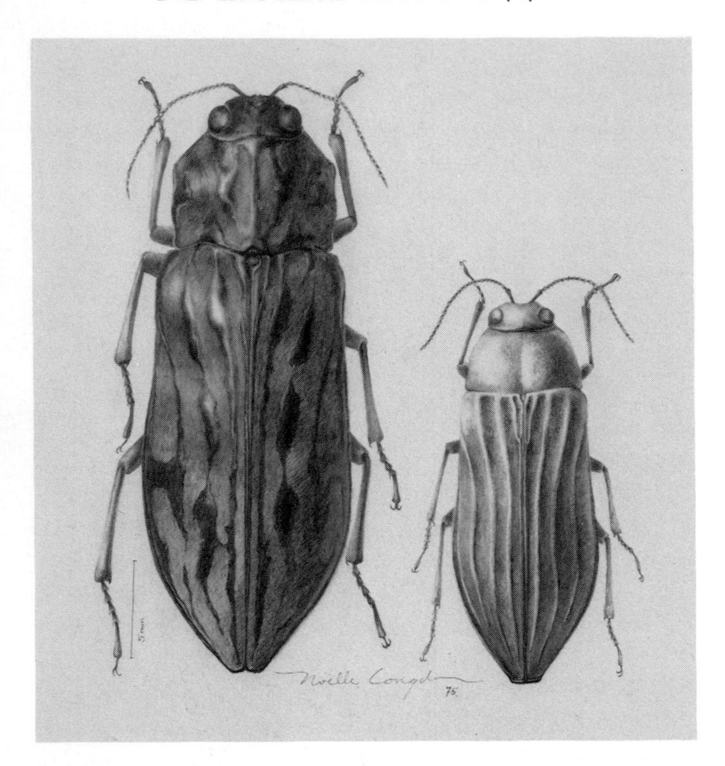

2-13. Nöelle Congdon. Metallic wood-boring beetles. Graphite pencil on film.

2-14. Nancy Williams. Wash on cold-press illustration board.

The trained eye

Both scientist and artist gain by training their observation skills with an eye to accuracy, rhythm, and pattern. The artist will understand the subject more thoroughly, and the scientist will be able to study and interpret his subject with a more discriminating eye.

The professional scientific illustrator must of course be familiar with his subject in more than a visual way. He must be aware of the structure and function of its various parts and recognize the anomalies (individual variations) in his specimen and the differences between related species. In other words, he must see not just an individual specimen but one specimen in relation to many species in the broader context of his entire field. This knowledge is necessary in order to produce the most effective selective drawings. For this reason most scientific illustrators specialize in one field.

Representative and individual specimens

To decide whether to render a specimen with all its individual idiosyncracies and peculiarities or to draw a representative specimen with the typical common characteristics of its species, you should consider the purpose of the drawing. If the drawing is to illustrate a unique prototype, the specimen must be drawn with every eccentricity in place. Usually, however, you are drawing a typical specimen and thus have the option of adding or straightening hairs, removing worm holes, repairing cracks and broken pieces, and moving or removing leaves and branches. You may use many specimens to produce one drawing based on their corporate characteristics, but take care not to misinterpret them while combining, simplifying, or synthesizing their elements. Never attempt to draw such a perfect and neat specimen that it loses its naturalness. None of nature is "perfect," and to draw it perfectly denies its inherent nature (2-15).

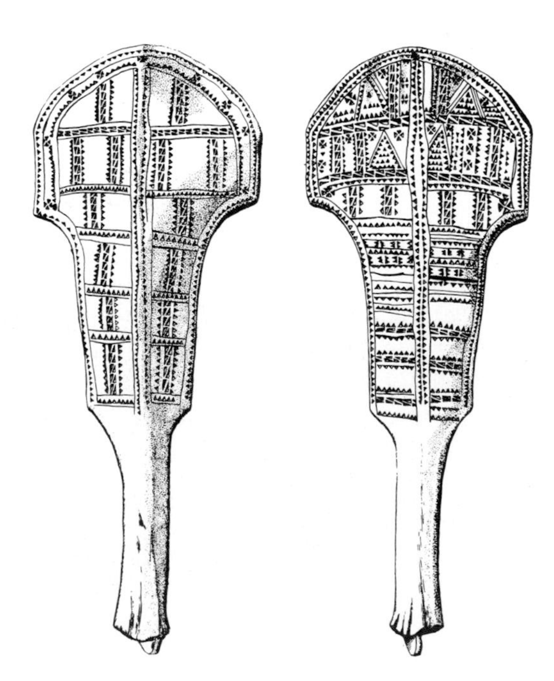

2-15. Colleen Hill. Indian war paddle. Ink on paper.

POSITION

Now that you have acquainted yourself thoroughly with your specimen, the first decision to make is the position in which to draw it. Do you want to show it at an oblique angle (2-16), or will a straight-on view (2-17) exhibit its shape best? When you want to communicate specific information to an audience, you must draw the subject in a straightforward manner with no tricky angles to puzzle the viewer. Even an accurate drawing in an unfamiliar or confusing position can mislead or create an illusion. Avoid the "different for difference's sake" and the ambiguous view. Do not assume, however, that the traditional view is always best. A dramatic or unusual position might show the subject just as well and also be more interesting to the viewer. The creativity of the artist should be combined with the scholarliness of the scientist.

When you have decided on the most informative and recognizable view, place your specimen in front of you on your drawing board in a stable position. Fix it in place with plastic clay or wet, crumpled paper towels.

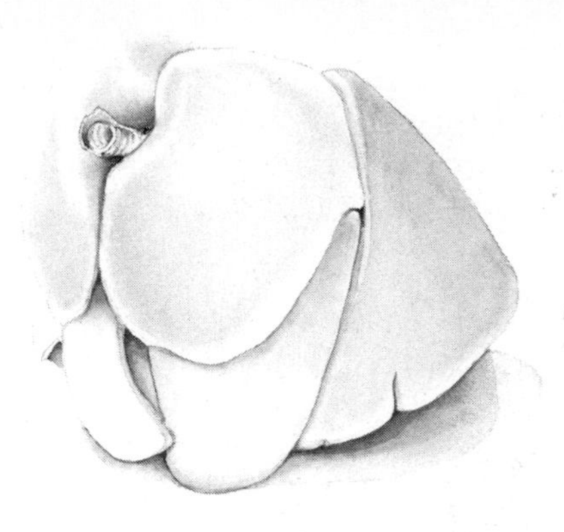

2-16. Katherine Miller. Monkey lung. Wash on cold-press illustration board.

SIZE

The next decision concerns the size in which to draw your subject. Start with an ordinary 8 1/2"-X-11" (22 cm-X-28 cm) sheet of paper and decide on a comfortable size for the subject. If you wish to include a great deal of detail, it will be easier to enlarge the subject two, three, or more times (2-18). If the subject is very simple and rather large, it may be easier to draw it in reduced size (2-19). Intermediate subjects are best rendered in actual size (2-20).

Measuring is easiest if the drawing is the same size as the subject (one-to-one). The next easiest measurements are simple increments such as half or twice the actual size. Other variations are limitless.

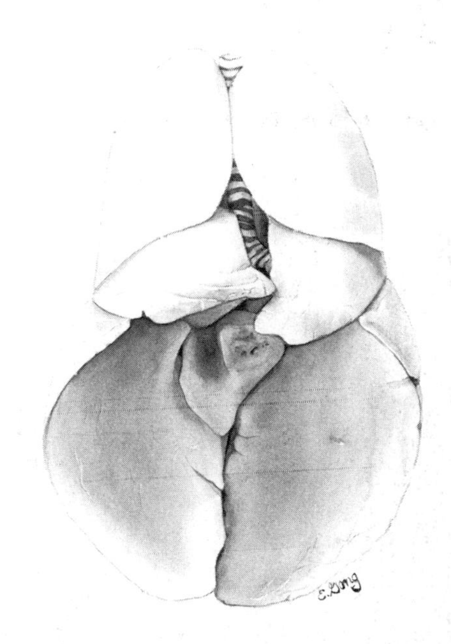

2-17. Elizabeth Gong. Monkey lung. Wash on cold-press illustration board.

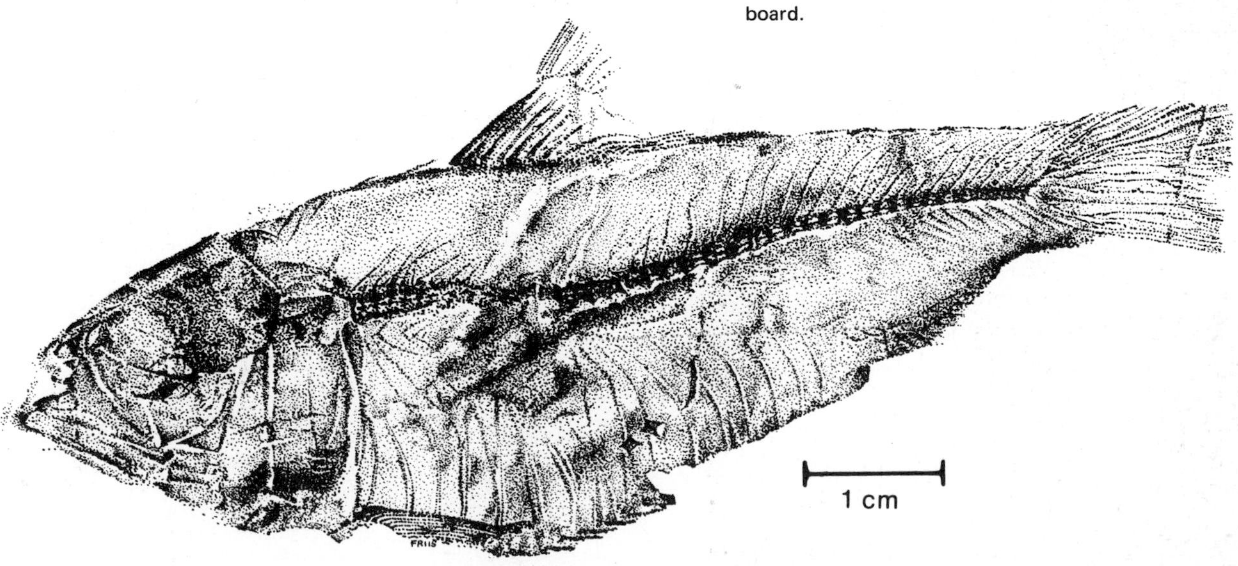

2-18. Anton Friis. Petrified fish. Ink on paper.

2-19. Craig Staude. Seaweed. Ink on paper.

2-20. Margaret Watson. Holly. Ink on paper.

Unless drawing a familiar subject, the artist must give the observer some idea of its relative size and bulk. The drawing may be the same size as the subject, or it may include a scale or code relating the size of the drawing to that of the subject (2-21a,b). If you are sure of the final published size of the drawing, you may write the scale as a fraction (2-21c). The bulk of the subject may be represented by shading and cast shadows, giving it a third dimension. It may sometimes be necessary to draw several views of the subject in order to show this (2-22).

Printed size

Drawing size also depends on the size in which it is to be printed. Measuring the page and column length and width and noting the usual style of the proposed publication can help you to make that decision. A drawing is always rendered larger than it is to be printed, as the reduction process improves and sharpens its quality. If there is a good deal of detail to be included, work the drawing two or three times larger than it is to be reproduced, which makes it easier to render the small detail. If the subject is fairly simple, you can draw it one and a half times the printed size, and it will reduce well. At the beginning of a project have the first drawings reduced to publication size on a photocopy machine to see whether line quality, detail, and overall comprehension are retained (2-23).

Series of drawings

When a number of related drawings are to be printed in the same publication, draw them the same size so that they can be reduced by the same amount, regardless of the size of the specimens. For instance, if you are drawing a number of specimens that vary in length from 1″ to 15″ and all are going to be printed at column width (3″), make all the drawings 6″ wide. If all are drawn the same size and reduced the same amount, the rendering in the publication will be uniform. Each drawing should be accompanied by a millimeter scale informing the reader of the actual size of the subject (2-24).

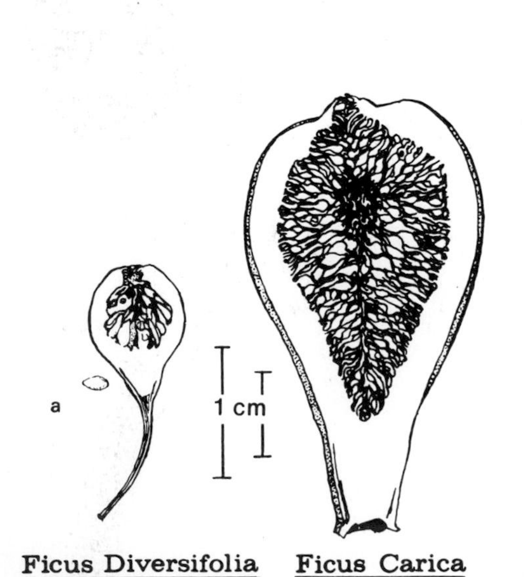

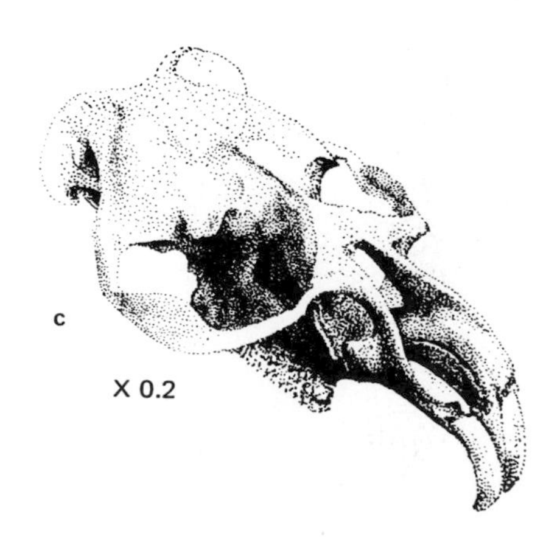

2-21. a. Kathleen Skeele. Fig. Ink on paper. b. Juanita Janick. Fir. Ink on paper. c. Daniel Cook. Mountain beaver skull. Ink on paper.

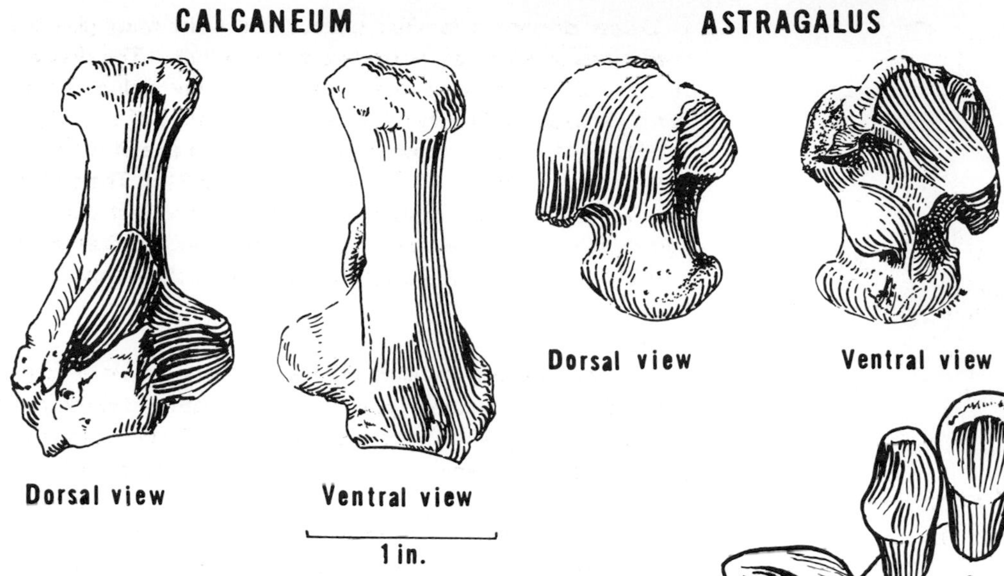
CALCANEUM
ASTRAGALUS
Dorsal view
Ventral view
Dorsal view
Ventral view
1 in.

2-22. Beverly Witte. *Hoplophoneus.* Ink on paper.

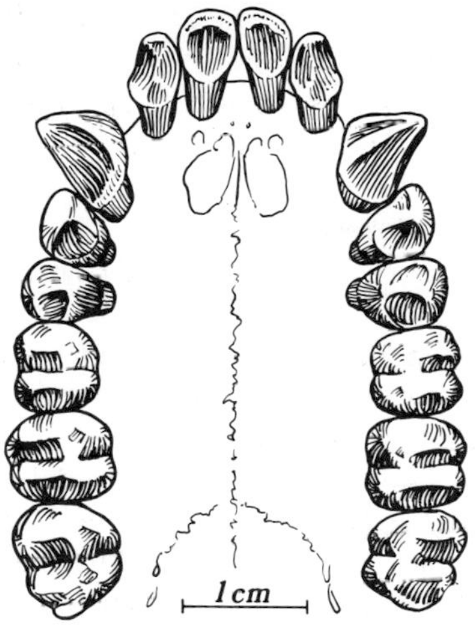
1cm

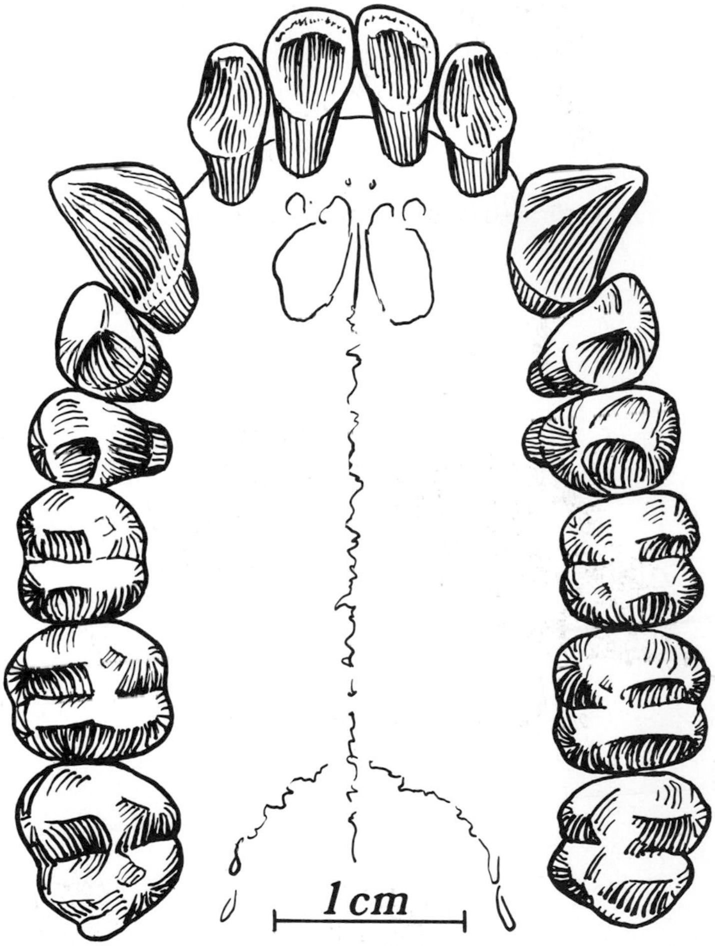
1cm

2-23. Phyllis Wood. *Colobus polykomos* (black-and-white monkey). Ink on scratchboard.

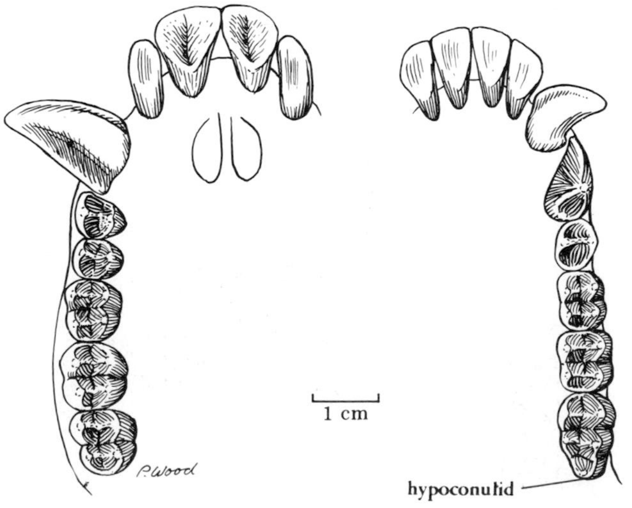
1 cm
P.Wood
hypoconulid

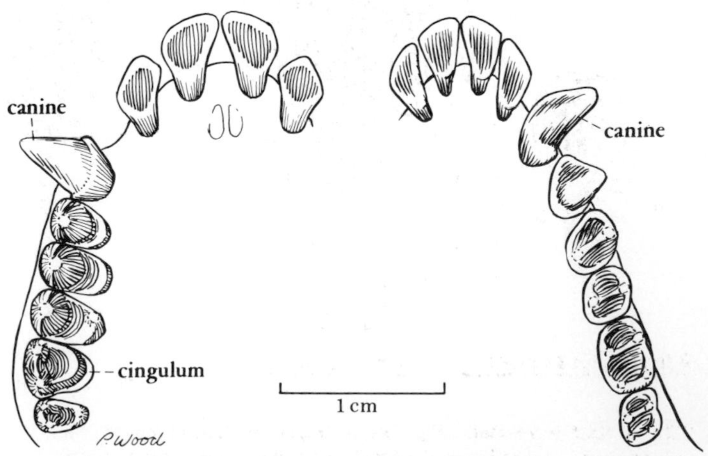
canine
canine
cingulum
1 cm
P.Wood

2-24. *Macaca nigra* (Celebes ape), *Saguinus geoffroyi* (Marmoset). From Daris Swindler. *Dentition of Living Primates.* Ink on scratchboard.

INTERPRETING THE SUBJECT

There are four factors that shape the interpretation of a specimen after you have decided on its position and size. They are contour, texture, color, and detail.

Contour

The contour or essential shape is the first concern in drawing any subject. Details are built on the foundation of an accurate drawing of the basic form (2-25).

2-25. Patricia Veno. Skull. Ink on paper.

Texture

A thorough study and sensitive rendering of the surface textures are subordinate to but part of the shape of the main structure. The texture can be shiny or dull, soft or hard, smooth or rough, grooved or serrated, hairy or holey, wet or dry, or patterned with hexagons, pentagrams, or venation. The texture is rarely the same overall: it is usually a varied combination (2-26).

2-26. Gail Peck. Romaine lettuce. Ink on scratchboard.

Color

Color is also subordinate to contour. In order to interpret drawings done in only black, white, and gray values, the artist should imagine the way in which a black-and-white photograph would depict the colors of the specimen and proceed from there. Think of color as a gray value painted on the surface. It takes on the contours of the main structure and of the surface structure (texture). It must not have more thrust and importance than does the contour of the subject (2-27).

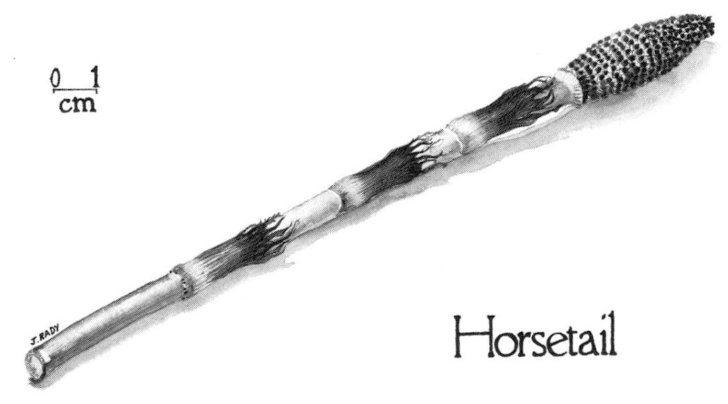

2-27. Jane Rady. Wash on cold-press illustration board.

Details

After firmly establishing the main contours and the surface pattern, texture, and color look with a critical eye at the myriad detail in your specimen. Inclusion and careful and selective interpretation of this detail often lends authenticity to your drawing. Determine how much detail is necessary to establish the integrity of the drawing and how much should be left out for the sake of aesthetics (2-28).

Measuring

Measure the width and the height of the specimen on the picture plane. Lightly draw a rectangle using the same measurements or a previously determined increase or decrease in scale (2-29). When measuring, the ruler or divider must always be at the same distance from the specimen. The measuring device must be at a right angle to the line of sight and parallel to the picture plane. You do not measure any part directly unless that part is parallel to the picture plane. Parts are measured as they appear on the picture plane (1-11).

Monocular measuring

It is necessary to make these preliminary measurements with one eye closed. Measurements are always made monocularly—with the same eye. Binocular measurements result in changing relationships.

To demonstrate this principle, hold your ruler 1" from your specimen and measure your specimen using only your right eye (2-30). Now close your right eye and measure your specimen using only your left eye (2-31). Notice how all the relationships change. Now try to measure with both eyes open. Holding your head still and viewing with both eyes is seeing in stereo, and what you see is binocular space. Classical perspective is the result of holding your head still and looking with one eye: what you see is monocular space.

Although you must hold your head still and make monocular measurements, you must still use your three-dimensional sense in interpreting the subject. If you follow these measuring rules inflexibly, you may end up with a flat drawing resembling a traced photograph.

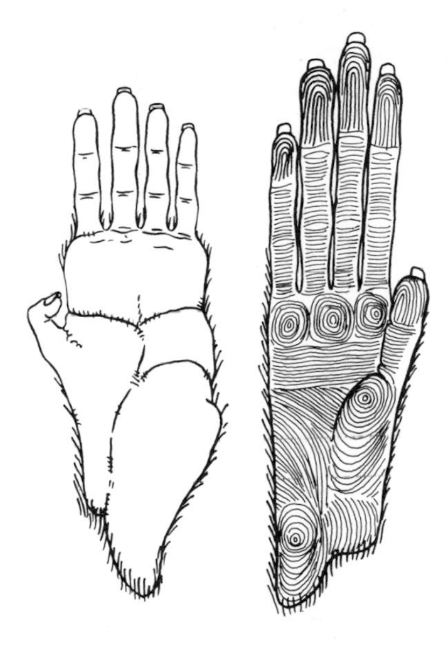

2-28. Phyllis Wood. Hands of *Erythrocebus patas* (red monkey) and *Macaca nemestrina* (pigtail monkey). Ink on paper.

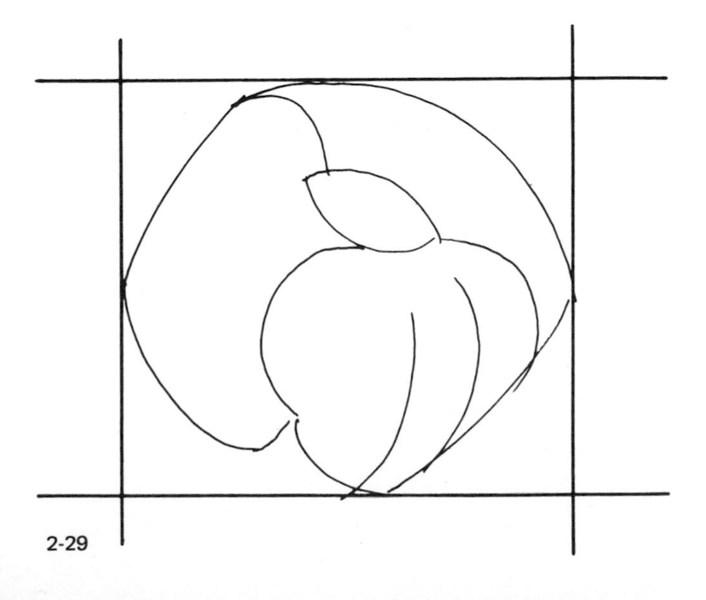

2-29

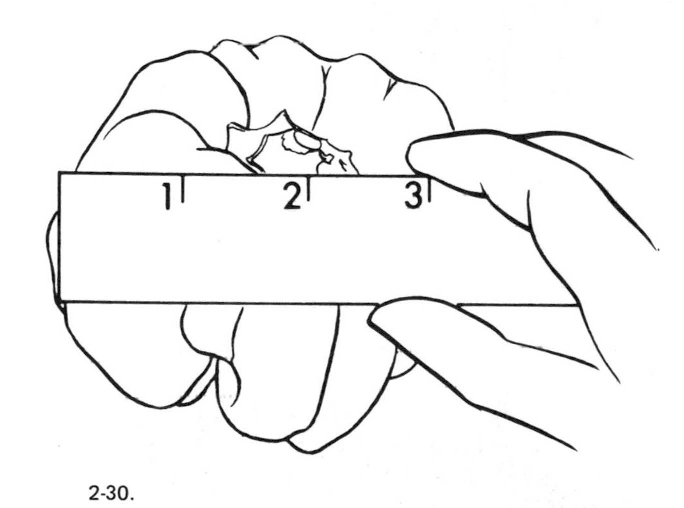

2-30.

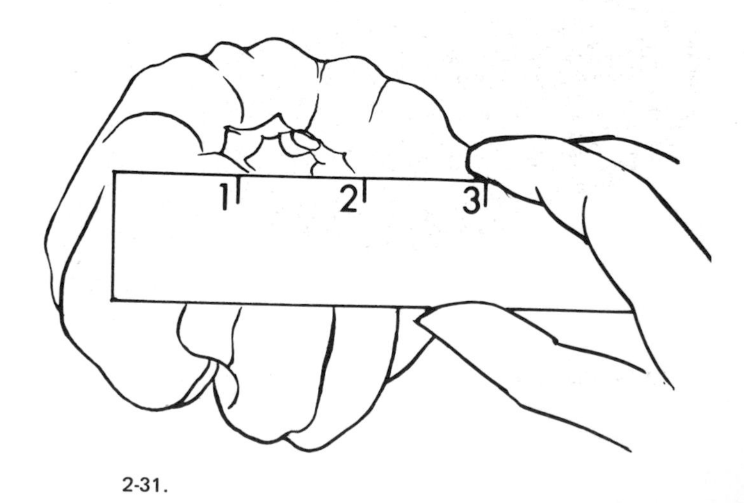

2-31.

Parallax

If your eye is too close to the specimen, the parallax phenomenon may seem to distort it. Parallax is a kind of extreme perspective that produces an exaggerated fish-eye view. The front and center of the subject are large compared with the peripheral parts; less of the sides is visible (2-32). The further the eye is from the subject, the less obvious is the parallax effect. All parts of the specimen are in a more correct relation to each other, and more of the sides is visible (2-33).

In drawing some subjects, such as a long stem or bone, the eye must view the specimen from a moving site along the long axis. If viewed from a constant point at the middle of the subject, both ends of the specimen would converge to two vanishing points in opposite directions due to the effects of perspective. This would produce an awkward if not violent drawing. Photographers are familiar with this phenomenon and have lens capabilities to correct for parallax. They occasionally use parallax on purpose to produce a fish-eye distortion (2-34).

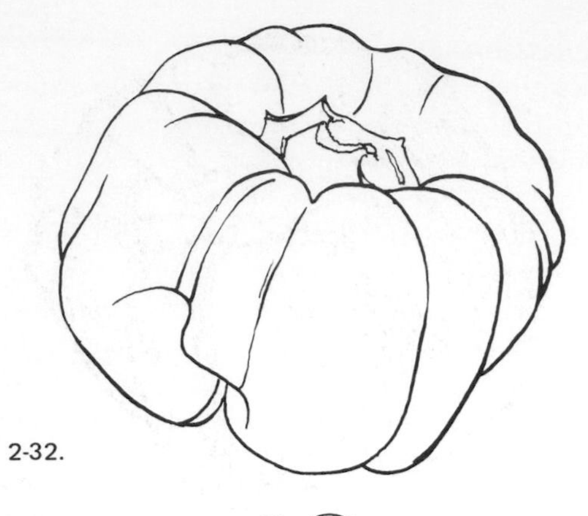

2-32.

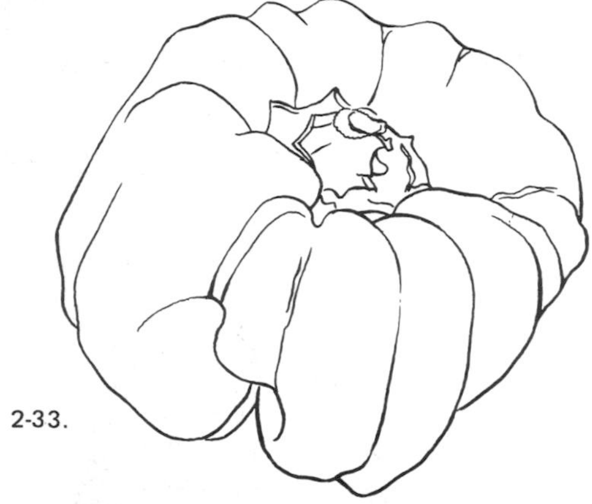

2-33.

2-34. Michael Sweeney. Reservoir, Rochester, N. Y.

Limits

The drawing of the specimen should touch the measured rectangular outline on all four sides. All the measurements made of the specimen are in relation to the outline of the rectangle. A landmark may be 1″ from the left side, halfway from the top, or one-fourth of the way down, for example (2-35). Check these measurements again and again to make sure that nothing has "slipped." Study the negative space (between the specimen and the outline of the rectangle) as well as the positive space that is occupied by the specimen itself for accuracy (2-36). Test angles and relationships of angles against one another.

Flick your eye quickly back and forth from the subject to the drawing. It can recognize discrepancies between them. If you find that a portion of the drawing is slightly off, redraw it before erasing the incorrect part. Draw your lines lightly at the beginning, critically readjusting and correcting them and gradually sharpening them as you become sure of their accuracy. If a large part of the drawing is incorrect but you wish to preserve some of it, use a tracing-paper overlay for changes.

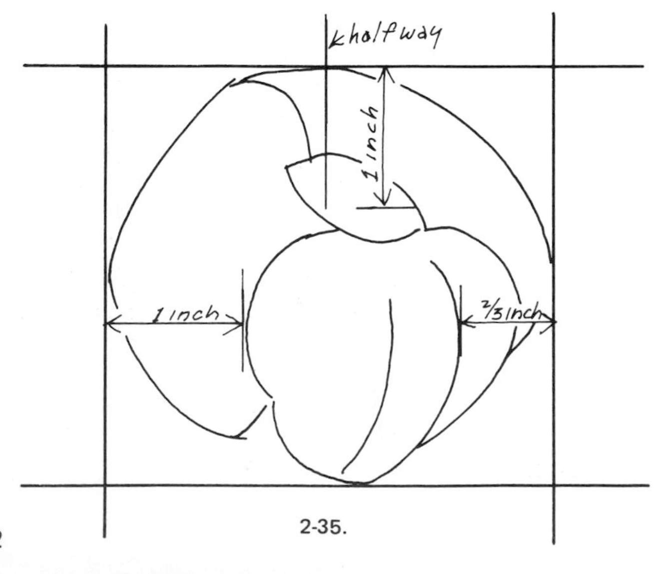

2-35.

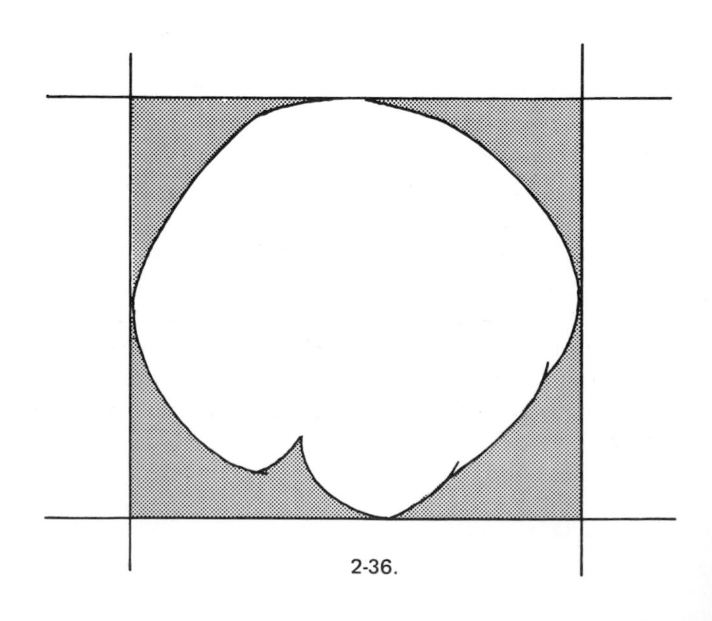

2-36.

Character

While making all these computations to ensure mathematical precision, the artist must also concentrate on recreating the character of the subject. It is helpful in drawing a green pepper, for instance, constantly to be thinking, "green pepper, green pepper, green, hard, shiny, tangy, crunchy, fleshy." After some practice you should be able to draw with precision without being aware of the underlying mechanics, as you are completely involved in experiencing the essential character of the subject. The result should be delicate without being timid and gutsy without being gross (2-37).

While you are completing your finished drawing, always keep your specimen in sight and refer to it often. Do not depend entirely on the preliminary drawing. As accurate and complete as you may think that it is, you should continue correcting, revising, and perfecting as you render the finished drawing. Continuous questioning of your judgment will be rewarded by an honest replication.

In this chapter we have referred mainly to a stable subject that does not change during the drawing process. Live botanicals, for example, change rapidly while you are drawing them, either by wilting or developing or turning toward the light. Microscopic specimens slowly change and divide. The artist must decide on the position and scope of his drawing at the beginning of his project and then adhere to that decision, using the specimen as a reference point.

2-37. Jane Rady and Phyllis Wood. Green pepper.

Measuring devices

There are various devices for measuring a specimen accurately. One, of course, is the eye. While mechanical devices calculate precisely, the eye is attached to the brain, which can create order in the drawing and thus control and make sense of the measurements. The eye and the brain should always be used in combination with any other device.

A plastic see-through ruler is the simplest and often most satisfactory device. Another type of mechanical device is a plain divider, which is more accurate than a ruler. One must take care not to measure with them at an angle but always on the picture plane. They can be used to expand or to reduce a subject by using the diagram (2-38).

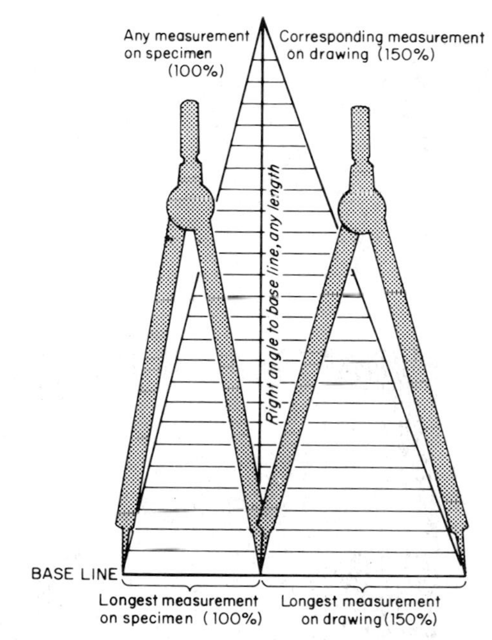

2-38.

Every scientific illustrator who works with small macroscopic subjects should have proportional dividers, which are double-ended pointers with an adjustable pivotal screw that can be set at different ratios. While the pointers at one end measure the subject, the pointers at the other end automatically open to the correct measurement for the drawing at the selected ratio. This ratio is adjustable for reduction or enlargement in any amount up to the capacity of the dividers (2-39). In the drawing the pivot is set so that the short pointers consistently measure two-thirds the measurement of the long pointers.

A device called a Camera Lucida or a Leitz drawing tube can be attached to a dissecting microscope and adjusted to various enlargements. While one eye looks in the scope, the other eye "sees" the image on the paper.

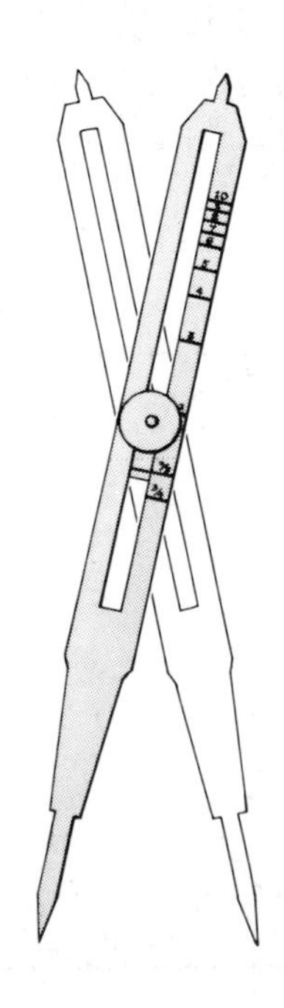

2-39.

The eyepiece of a microscope can be fitted with a grid. The view of the specimen in the scope grid can then be transferred by drawing it on grid paper, effectively enlarging the thin section on the slide to any desired size.

A machine commonly called a "lazy lucy" projects the image of an object onto translucent or opaque paper. The projection can be made in a wide range of reductions, enlargements, or the same size. It is done with strong lights, lenses, and sometimes mirrors. These devices should be used critically, as in any projection there is some distortion. Slides can also be projected on the wall and traced as a basis for a drawing.

TRANSFERRING THE DRAWING

When the preliminary drawing is "perfect," it should be transferred to the paper or board on which you are going to do the finished drawing. There are several methods of doing this, each with its own advantages and disadvantages.

Double transfer

For a drawing with a great deal of detail or sensitivity double transferring is the preferred method, as it transfers all the spontaneity of your original drawing and at the same time preserves it. An HB or softer pencil must be used on your preliminary drawing: if it is done with a hard pencil, it will not transfer.

Tape a single piece of tracing paper securely over the top of the drawing and place it on a hard, smooth surface. Burnish the back of the tracing paper thoroughly in several directions with a burnisher, a scalpel or spoon handle, or your thumbnail. Lift the tracing paper up partially and check to see whether the entire drawing has been transferred. When all has been transferred to the tracing paper, tape the transferred drawing face down on the paper or board chosen for the final drawing. Repeat the burnishing on the back of the tracing paper until the complete drawing is transferred to the final ground. You may need to strengthen some of the lines at this time and to remove any bits of pencil dust gently with a kneaded eraser, as they may interfere with the final rendering (2-40).

Single transfer

The principle of the single-transfer method is the same as that of carbon paper, except that nonindelible graphite is the transfer medium instead of indelible carbon. The simplest method is to darken the back of the preliminary drawing with a medium-soft pencil. Tape it in place on the final drawing paper. Trace the preliminary drawing directly onto the final paper with a hard, sharp pencil (4H). If you wish to preserve your original drawing, you can use purchased graphite paper (Saral brand) or make your own from tracing paper. The final drawing paper should be on the bottom, followed by the Saral, the preliminary drawing and a clean sheet of tracing paper. Trace through all three layers (2-41).

Projection

A lazy lucy can project your drawing either onto your final drawing paper or through translucent drawing paper, depending on which kind you have. While some distortion is created by this method and it has the disadvantage of dulling the freshness of your first drawing, you can easily transfer while reducing or enlarging your preliminary drawing.

Direct tracing

If you are using Mylar or some kind of acetate film or vellum that is fairly translucent, you can either pencil-trace your final drawing directly or actually render the drawing by tracing it directly. It is preferable to do most of your rendering after removing the preliminary drawing from under the film, as the image underneath interferes with the building of values in the finished drawing. If you use drawing papers that are less translucent, you may choose the direct-tracing method by employing a light box.

Reducing or enlarging

You may find that the drawing size that initially seemed comfortable and correct is not the right size for your layout and that your drawing must be larger or smaller. You can place a grid over the drawing and then transfer it onto a grid in another scale (2-42). There are also several ways in which the scale can be changed mechanically. Some copying machines can reduce to specific levels. Photocopies, stats, or photographs can be made in almost any size. A lazy lucy can project a drawing within a large size range. There is, of course, a cost factor and also a convenience and time factor attached to any of these methods. You can also change the scale by redrawing it by eye. The disadvantage of this method is that you are doing the work twice, and the artist, like any other worker, should use the easiest and least time-consuming methods at his disposal. There is nothing artistically unethical about using mechanical methods.

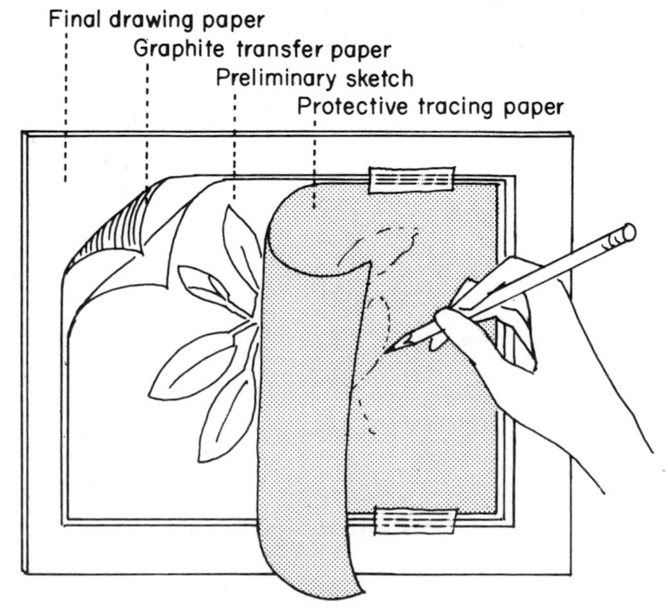

2-41.

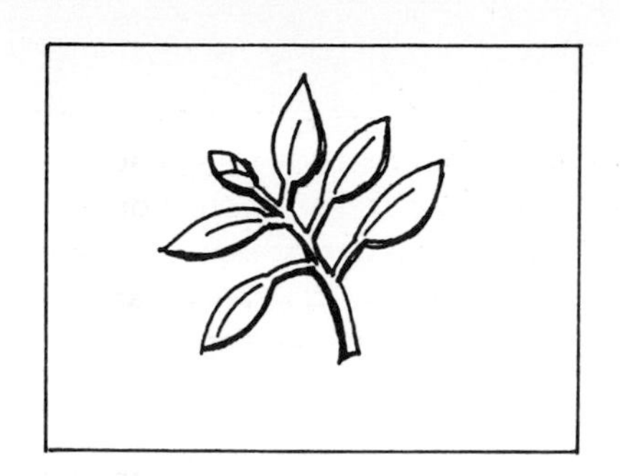

1. Preliminary sketch

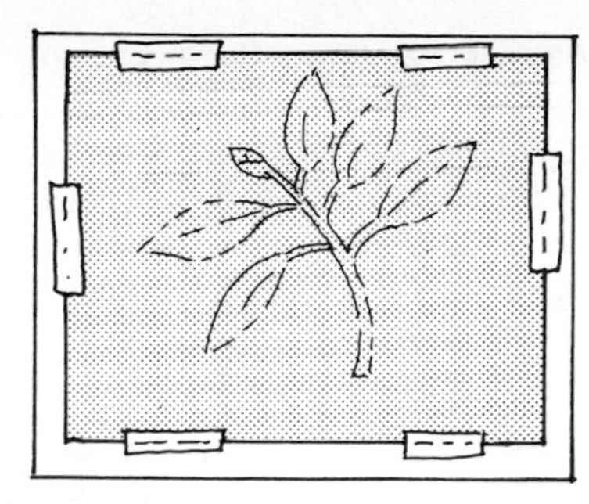

2. Tape tracing paper to sketch

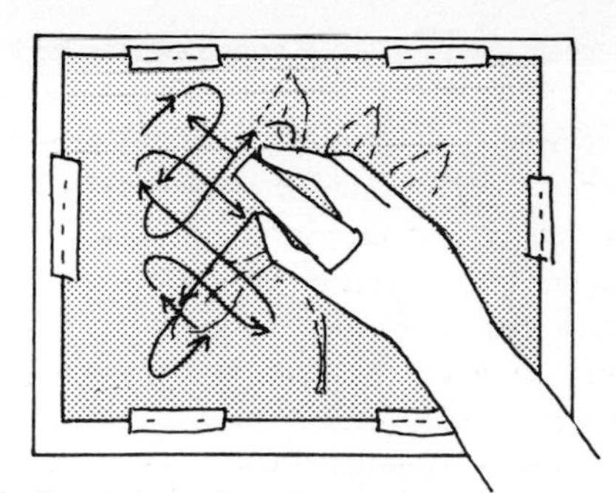

3. Burnish on hard smooth surface to transfer sketch to tracing paper

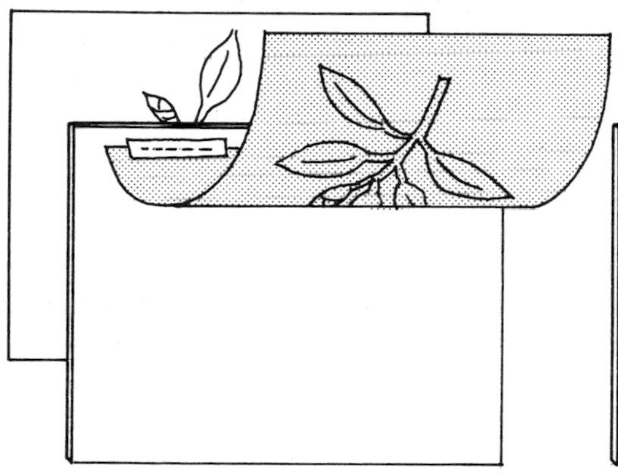

4. Tape tracing paper to final drawing paper

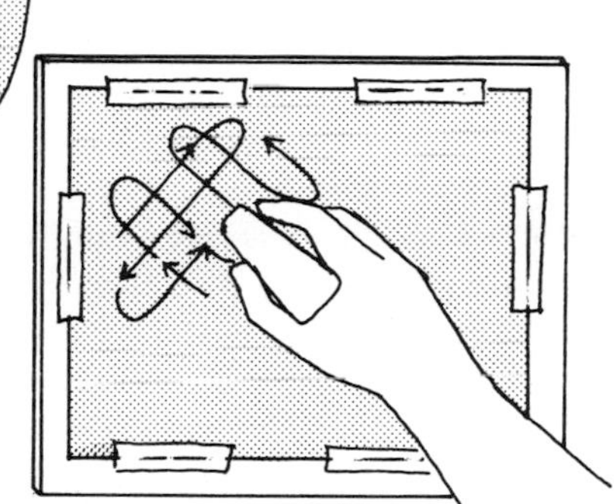

5. Burnish to transfer sketch to drawing paper

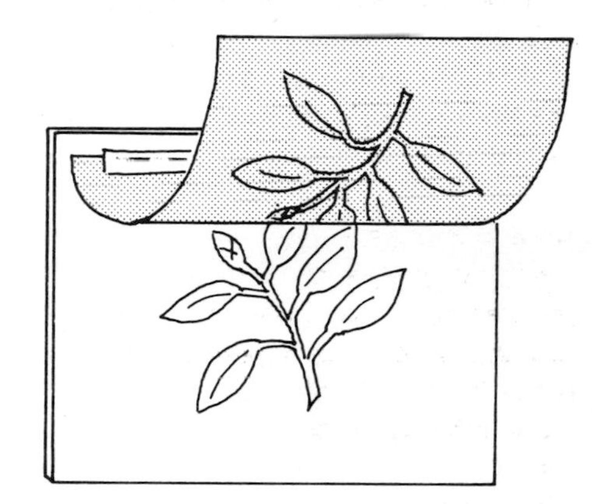

6. Check for complete transfer to final drawing paper

2-40. Kathleen Todd and Phyllis Wood.

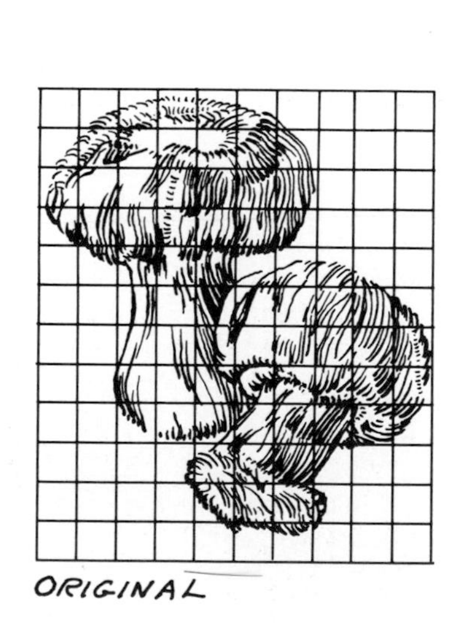

150%

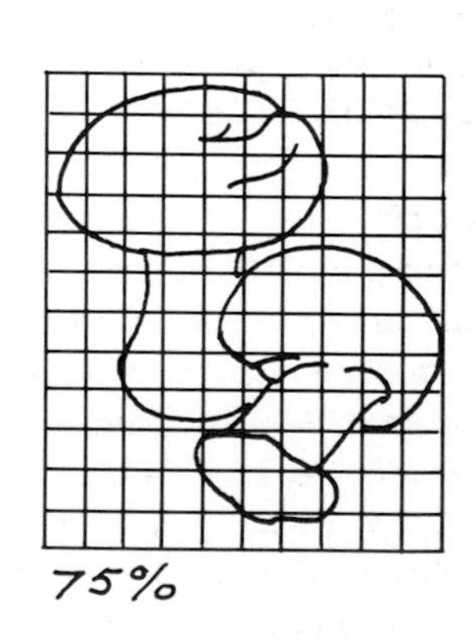

2-42. Sharon Feder and Phyllis Wood. Mushroom.

CHAPTER 3.

Light and Shadow

To separate the subjects drawing and light and shadow into two chapters is artificial and arbitrary, since the two are inextricably joined. Light-and-shadow relationships define any shape. If the highlights and shadows were removed from a subject, all that would remain would be variously shaped areas of flat color (3-1).

Every shadow, highlight, and cast shadow in a drawing should be present only for the express purpose of defining the shape of the subject. If a light or shadow does not help to define the shape, it should be altered, omitted, or added to so that it does. The artist should be able to conceptualize ideal lighting in interpreting the lights and shadows on a subject's shape.

3-1. Phyllis Wood. Herring gull. Wash on cold-press illustration board.

VALUE RELATIONSHIPS

Light and shadow are interpreted by value relationships that define a subject's shape. The value of a surface is either light or dark, depending on how directly or indirectly the light rays strike it. The lightest surface or highlight area is caused by a single light source that strikes it at a right angle. The darkest surface, which would appear on the opposite side, receives no light rays. All the intermediate surfaces vary according to the angle at which the light rays strike them (3-2).

3-2. Pencil on paper.

Reflected light

The presence of a single light source is rare. In addition to the primary source of the light and shadows that define shapes both reflected light from other surfaces and diffused light surround us. Almost every surface has a reflective quality that bounces light onto surrounding surfaces. The lighter and shinier the surface, the more reflective it is.

If you place a sphere or a spherical object on a dull black surface, you will notice that no light is reflected from the surface to the sphere (3-3). If you place the same sphere on a white surface, light is reflected upward from the white surface and the sphere is defined more clearly. Our eyes expect light to be reflected from the white surface to the darkest shadow area of the sphere: reflected light therefore helps to define the shape of the sphere or of any subject. If you place the subject that you are drawing on a white surface, you will see it better, interpret it better, and therefore draw it better.

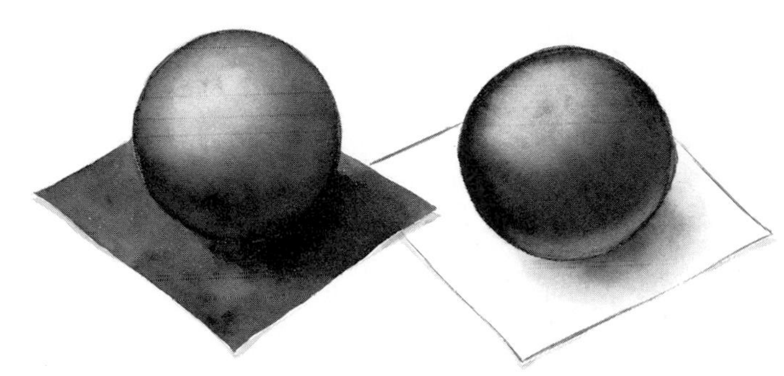

3-3. a. Wash on cold-press illustration board.

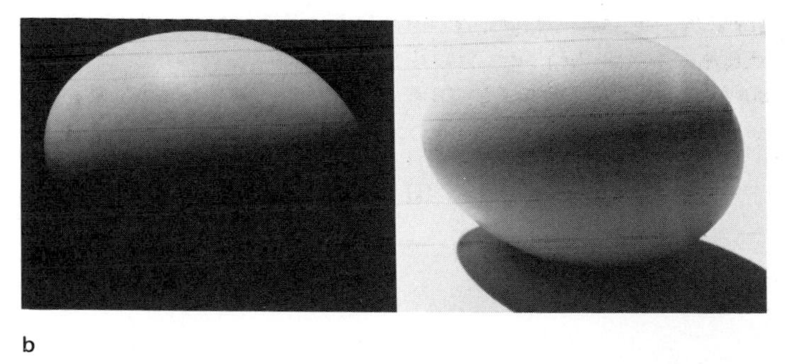

b

Basic values

The sphere is an ideal model for studying the four basic value areas of light and shadow. When you understand the principle by which these areas are distinguished, you can apply it to any other shape. The four areas are: the highlight, the intermediate shadow, the core dark, and the reflected light (3-4). Light can also be reflected inside a subject (3-5).

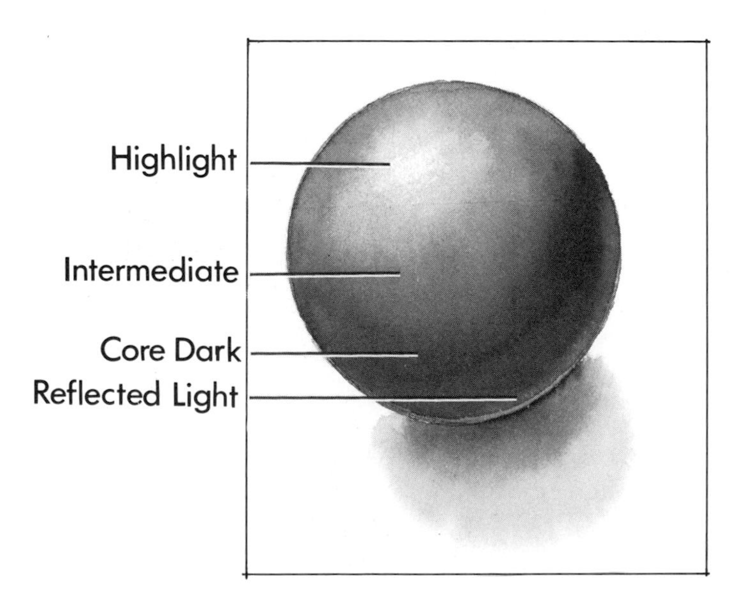

3-4. Wash on cold-press illustration board.

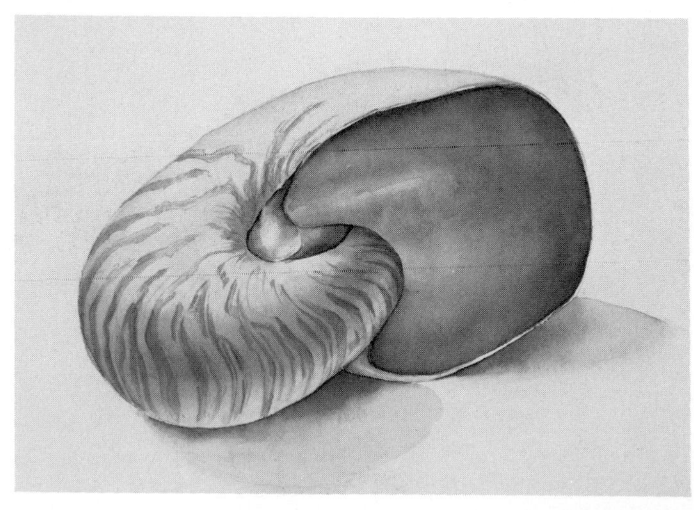

3-5. Marjorie Davis. Shell. Wash on cold-press illustration board.

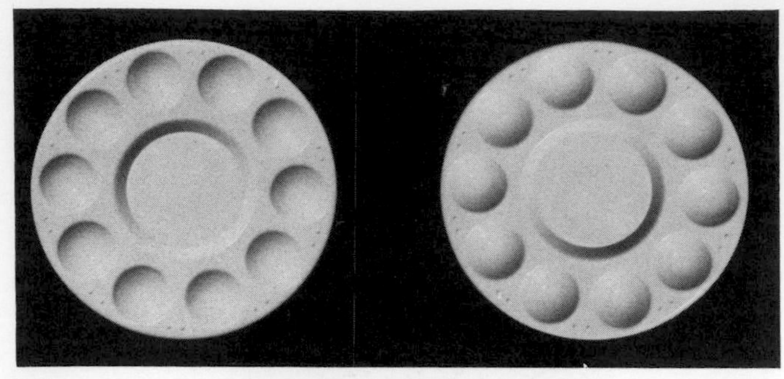

3-6.

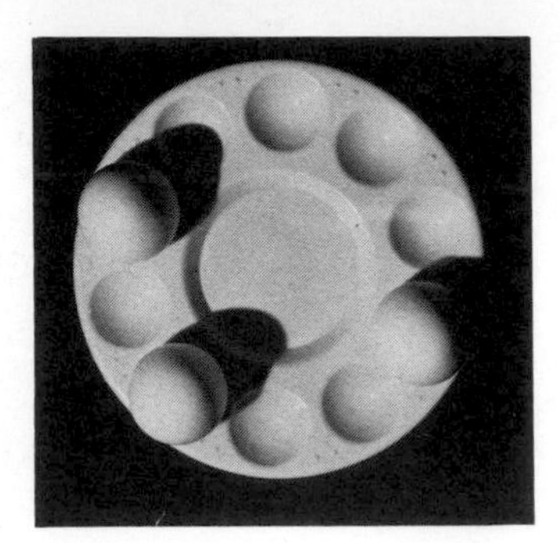

3-7.

Ideal light

The ideal setup for your specimen includes light that is steady and consistent in quality and direction at all times. Drawing areas do not always have ideal conditions, however, and even the best natural north light gradually changes during the day and disappears in the evening. The scientific artist cannot always draw shadows as he sees them. He must be able to interpret them as if the subject were under ideal and unchanging lighting conditions. He should make his shading plan early in the drawing stages and then stick to it, no matter how much the actual lighting changes.

Shading conventions

In scientific illustration the convention of the upper-left light source is traditional because it is the most satisfactory direction for a right-handed artist. With a consistent light source the viewer can most easily interpret the drawing. He is conditioned to differentiate between a convex and a concave surface.

In the left-hand picture of the paint pan (3-6) the cups are instantly recognized as depressions because the viewer is accustomed to an upper-left light source. When the same picture is turned upside down, the cups look like bumps instead of depressions because the light source is from the lower left. We are so accustomed to the convention of upper-left lighting that any variation is deceptive. When additional clues such as eggs are added to the paint pan (3-7), the shapes are readily recognizable as depressions.

Use a consistent light source for a number of related drawings on a single page. It would be very confusing if the light source varied from one drawing to another. Different elements in a single plate are often drawn separately and then fitted together. The top-to-bottom orientation should be decided upon before the drawings are rendered so that they have a common light source (3-8).

In addition to the main light source and the reflected light the artist may add a secondary weaker light source from another direction to lend variety and to define the subject.

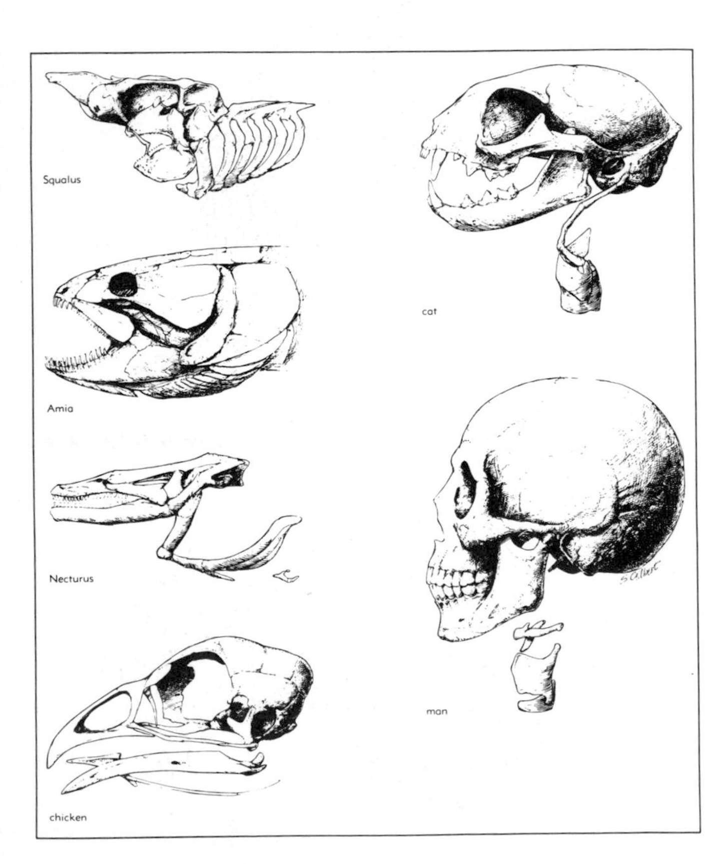

3-8. Stephen G. Gilbert.

Interpreting the specimen

The same four factors that govern the drawing of the specimen guide the value relationships of light and shadow. These are: contour, surface texture, surface color, and detail.

Contour

The light and shading of the subject's general contour are the foundation upon which the shading of the other elements is built. Surface texture, color, and detail are all subordinate to the main contour. Squint at the subject in order to screen out everything but the light and shading of the main contour shape. This should be established firmly before the other three factors are considered (3-9).

Surface texture

Surface texture or pattern is clearest and most easily defined in the intermediate-value area. In this area the light strikes the texture laterally, and any roughness exhibits maximum contrast between dark and light. In the highlight area the light tends to flood into the roughness, in effect flattening it. In the dark areas the roughness does not receive enough light for contrast (3-10).

Shiny surfaces are high in contrast. The lights and darks are closely juxtaposed, with fewer areas of gradual value gradation (3-11). Smooth surfaces show a gradual, even value gradation, with no sudden changes in value (3-12). Dull or fuzzy surfaces have no bright highlights or very dark areas (3-13). Wet surfaces have small, brilliant reflections in surprising places, sometimes deep inside a dark area or underneath on the shadowy side. They occur wherever moisture catches a ray of light (3-14).

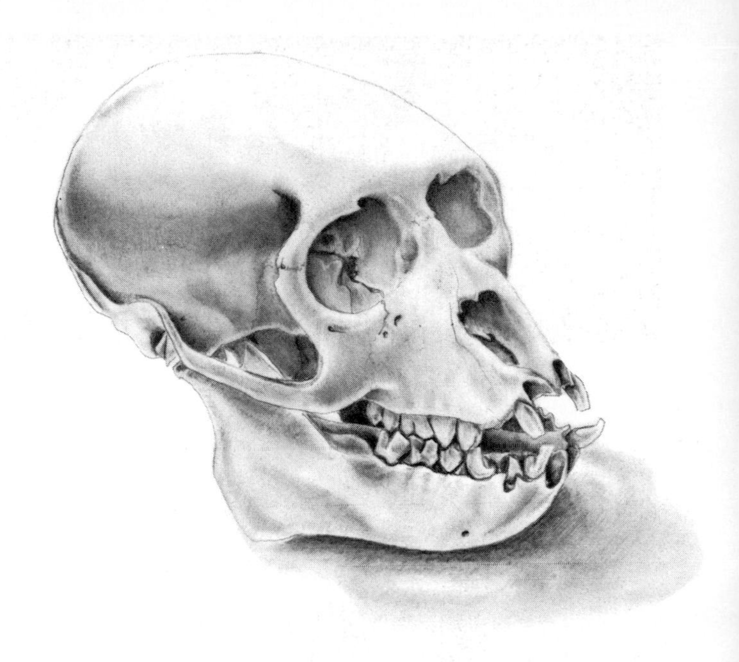

3-12. Katherine Miller. *Macaca nemestrina* skull. Graphite pencil on film.

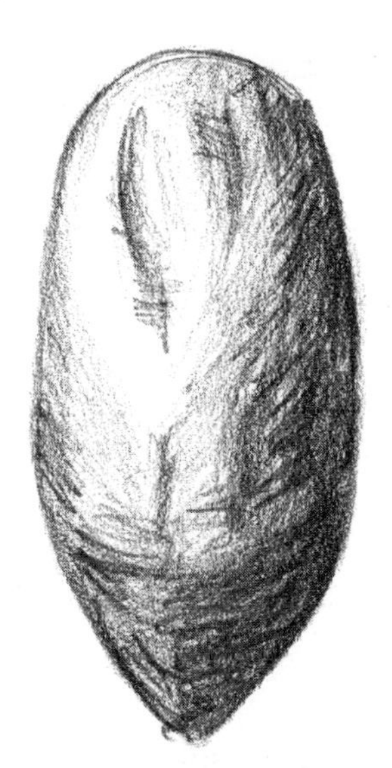

3-9. Phyllis Wood. *Cecropia* moth cocoon. Pencil on paper.

3-10. Nöelle Congdon. *Cecropia* moth. Carbon dust on ledger paper.

3-13. Nöelle Congdon. *Cecropia* moth. Carbon dust on ledger paper.

3-11. Marjorie Davis. Leaves. Wash on cold-press illustration board.

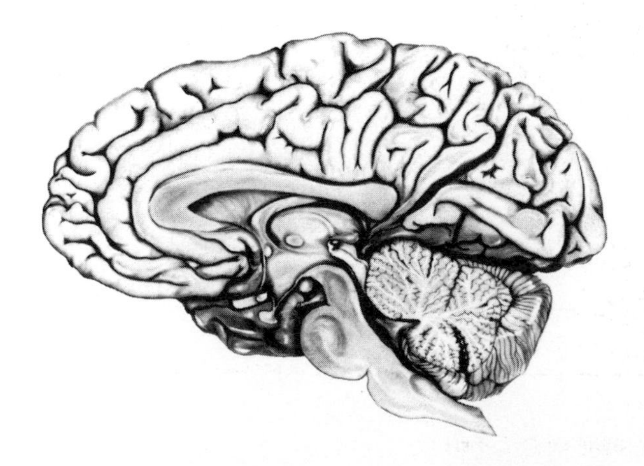

3-14. James Flaherty. *Homo sapiens* brain. Carbon dust on film.

Surface color

In a black-and-white or continuous-tone drawing the color on the surface of the specimen must be translated into corresponding values on the gray scale. It is again helpful here to squint at the subject. This will make the individual colors appear less significant, while their value relationships will become clearer. It is also helpful to imagine the manner in which a black-and-white photograph would interpret the colors on the gray scale (3-15).

Surface color is directly affected by the light and shadows that define the contour of the subject, becoming lighter in the light areas and darker in the dark areas. In a dark subject all values are relatively darker than in a light subject, but their value ranges may overlap. The highlight on the back of a black bird can be lighter than the deepest shadow on a white bird.

A color pattern is most clearly defined in the intermediate-value area. In the highlight area the bright light obliterates some of the color pattern, and in the dark shadows the values merge together (3-16).

3-15. Nöelle Congdon. Carbon dust on ledger paper.

Details

The delicate shadows and highlights that appear in small details help to define them and to give them authenticity.

The thickness of the leading edge of a shell or leaf should be shown by a shadow underneath it. This establishes the three-dimensional quality of the subject and also indicates which edge is positioned toward the viewer (3-17).

Even thin structures such as hairs and fibers have both a light and a shadow side. While it might be awkward to draw the length of each filament in three dimensions, it is very effective to suggest by means of an occasional shadow line that it does have a circumference (3-18).

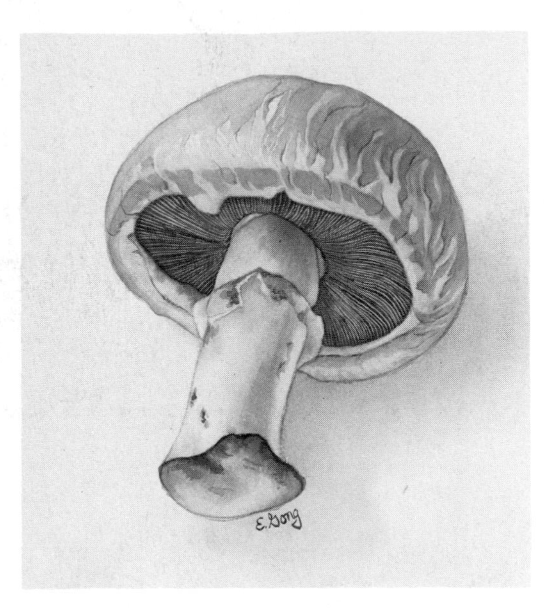

3-16. Elizabeth Gong. Mushroom. Wash on cold-press illustration board.

3-17. Nancy Williams. Wash on cold-press illustration board.

DOUGLAS FIR
(Pseudotsuga menziesii)

3-18. Juanita Janeck. Ink on paper.

A bump is described by lighting the near side and shadowing the far side, which merges with the cast shadow (3-19). Small holes or dimples are shown in just the opposite manner: the far side catches the light, and the shadow falls in the near side of the pit (3-20). Look for the lip edges that occur along overlapping structures, such as the bone above the molars. These catch the light and also cast a shadow.

ATMOSPHERIC PERSPECTIVE

The principle of atmospheric perspective is used in interpreting the range of values. Although this principle is not easily recognized in a small subject, it adds depth and a three-dimensional quality to the drawing (3-21). Imagine that the further parts of the specimen are covered with a thin film or veil of atmosphere. The focal point of the drawing, the foreground, exhibits crisper contrasts, darker darks, and lighter lights and is more defined than the background and more distant parts of the drawing. The highlights are confined to the foreground.

If some less important parts of the specimen also extend toward the viewer, think of the focal point of the drawing as in focus and the closer, less important parts of the drawing as out of focus and less clear.

CAST SHADOWS

The cast shadow helps to define the shape of the subject and establishes the subject in relation to another surface. It it not the same shape as the subject but a distortion of it (3-22). It is always subordinate to the subject. Because of the diffusion of light the cast shadow is not crisply defined. It has fuzzy edges, with the darkest part nearest to the subject and the lightest part on its periphery. It tends to merge with the shadow side of the subject (3-23).

3-19. Reed Eastman. Sea urchin. Wash on cold-press illustration board.

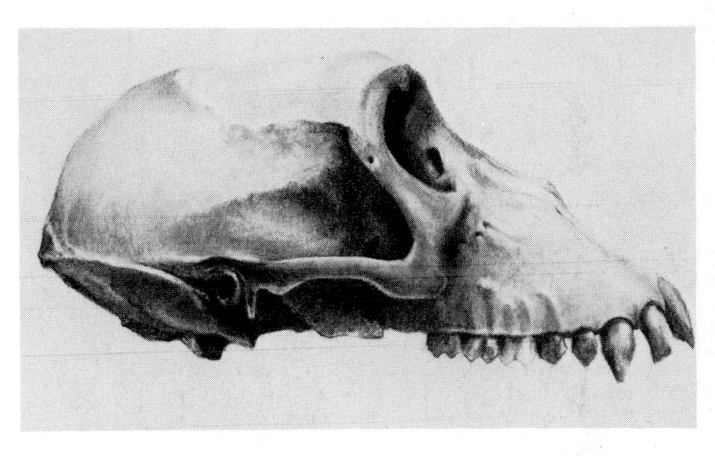

3-20. Daniel Cook. Rhesus monkey skull. Carbon dust on ledger paper.

3-21. Katherine Garnhart. Ink on paper.

3-22. Jane Rady. Wood. Carbon dust on video paper.

When interpreting the cast shadow in line shading, use passive parallel lines that follow the shape of the surface on which it falls. This may be either a flat surface or an adjoining variable shape. The shadow should not be exciting, since it is present only to define the subject, not as an entity in itself (3-24). A shadow on a flat surface is best defined by horizontal lines. Oblique or vertical lines, even if they follow the general shape of the shadow, seem to have a life of their own and should be avoided.

3-23. Dinah Wilson Stone. Shell. Wash on cold-press illustration board.

TRANSLUCENCE

Translucence is the characteristic of an object that can both be seen and seen through. Such objects include insect wings, watery subjects, eyes of animals, and bodies of various small forms of life, particularly small aquatic animals. The object itself, or top layer, is seen clearly, while the underlying or inside part is muted and shows less contrast. In rendering a continuous-tone drawing of a translucent subject it is usually best to do the underneath part first, keeping in mind that there is a covering layer (3-25). In a black-and-white line drawing you should do the top layer first, reserving space for the fainter rendering of the underlying structures (3-26).

A model for an eye or cyst can be found in a translucent drop of water, marble, or bead (3-27). In addition to the primary highlight, which falls where it would ordinarily on any sphere, there is a fainter secondary highlight at the opposite pole of the sphere. This secondary highlight is surrounded by the surface core dark. The inclusion of this secondary highlight gives a deep, "liquid" appearance (3-28).

Translucent subjects can be rendered very effectively on black or dark backgrounds, allowing the background to show through the translucent areas. In this case the translucent structures are usually lightest on the edges, where the subject is viewed obliquely; darkest, with the most background showing through, in the center, where you can see through the subject more easily (3-29).

3-24. Paula Richards. Ink on scratchboard.

3-25. Trudy Nicholson. Sailor-by-the-wind. Graphite pencil on frosted acetate.

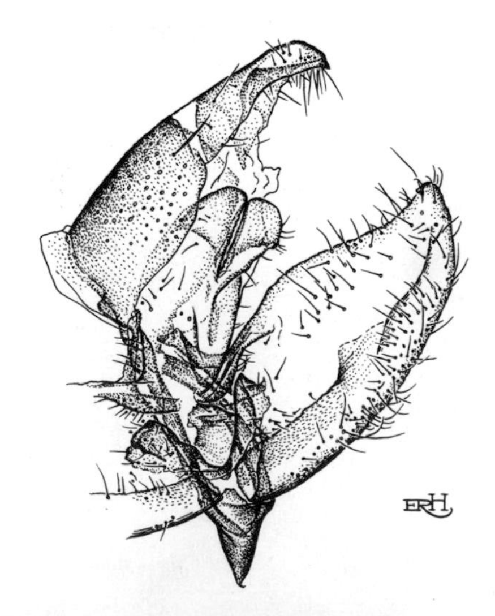

3-26. Elaine Hodges. *Cheimophilia rufocrinitalis* male genitalia. Ink on paper. From Ronald W. Hodges. *Moths of America North of Mexico.* Fascicle 6.2. (With permission of the Wedge Entomological Research Foundation.)

REFLECTIONS

Reflections occur on shiny metal, mirrored, or wet surfaces. If the reflective surface is horizontal, the reflection will appear as a distortion directly under the subject. The amount of distortion depends on the smoothness of the reflective surface, the angle at which it is viewed, and the reflective quality of the surface. Only in a highly polished mirror surface is the intensity of the reflection as clear as that of the subject itself (3-30).

OUTLINE DRAWING

Even an outline drawing should indicate the light source. Vary the weight of the outline, making it heavy on the shadow side and fine on the light side. This establishes the light source and helps define the shape and bulk of the subject. The outline on the shadow side should not be a static wire but the edge of the shadow itself (3-31).

b

3-27. a. Trudy Nicholson. Graphite pencil on frosted acetate.
b. Elizabeth Keohane. Wash on cold-press illustration board.

3-28. Phyllis Wood. *Nycticebus coucang* (slow loris). Ink on scratchboard.

3-29. Reed Eastman. Prismacolor pencil and backpainting on frosted acetate.

INTERPRETING PHOTOGRAPHIC SHADOWS

You may be called upon to draw either exclusively or partially from photographs, as they are sometimes the only available source material. This type of rendering forces you to translate unnatural shadows. If the photograph was taken with a flash, details in the shadows will be obliterated, the important contour shading lost, and the whole subject flattened. It is up to you, the artist, to reconstruct the three-dimensional quality of the original subject by inventing ideal lighting conditions (3-32).

3-30. Kathleen Todd. Apple. Color pencil on video paper.

a

3-31. Ink on paper.

b

3-32. a. Flash photograph. Infant *Macaca nemestrina* (pigtail monkey). b. Phyllis Wood. Wash on cold-press illustration board.

CHAPTER 4.

Black-and-White Drawing

Black-and-white drawing is the most useful technique for illustrating scientific publications (4-1). It is also the most difficult to master. According to Ernest W. Watson, former editor of *American Artist* magazine: "An artist reveals his strength or his weakness in a pen drawing. There is a merciless finality about a black line or spot which cannot hide under a camouflage of color or tonal charm. The test of an ink drawing lies in what is left out, fully as much as in what is put in."

Several considerations make black-and-white drawing the preferred technique.

1. It can be reproduced by any method and on any quality of paper.
2. It can be greatly reduced and still retain its integrity.
3. It is the least expensive to reproduce.
4. It is not dependent on the engraver for the quality of the screening process, which in an inexpensively produced journal may not be the finest.
5. Values from 100% black to 100% white are maintained from original to reproduction.
6. Delicate detailing is possible.
7. It can be done on either tracing paper or film, eliminating the transfer process.
8. It can be the fastest of all techniques to render.

With so many reasons to use black-and-white ink drawing, it is worth a good deal of patient practice to attain virtuosity in this technique.

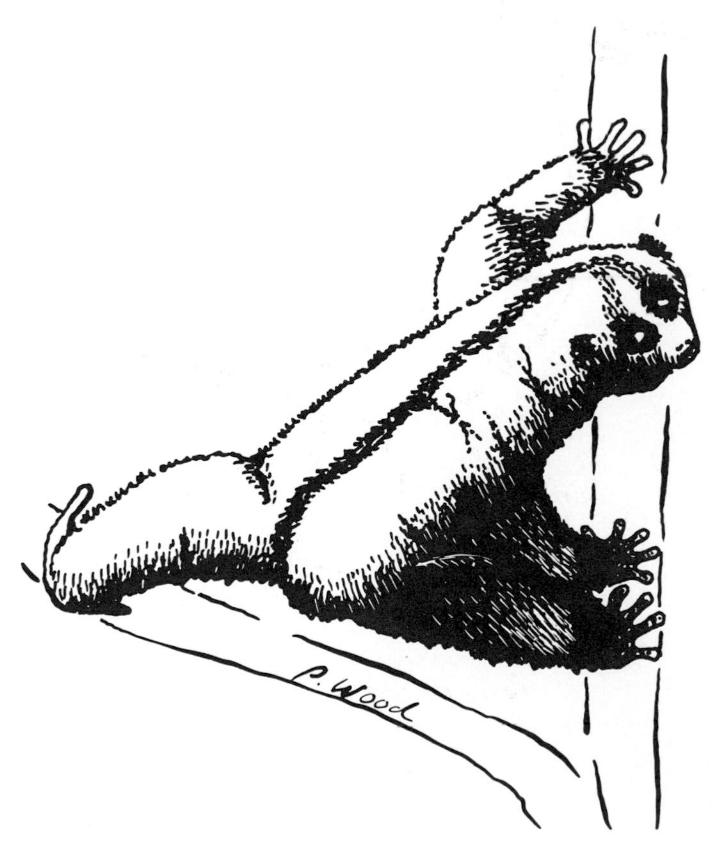

4-1. Phyllis Wood. *Nycticebus coucang* (slow loris). Ink on scratchboard.

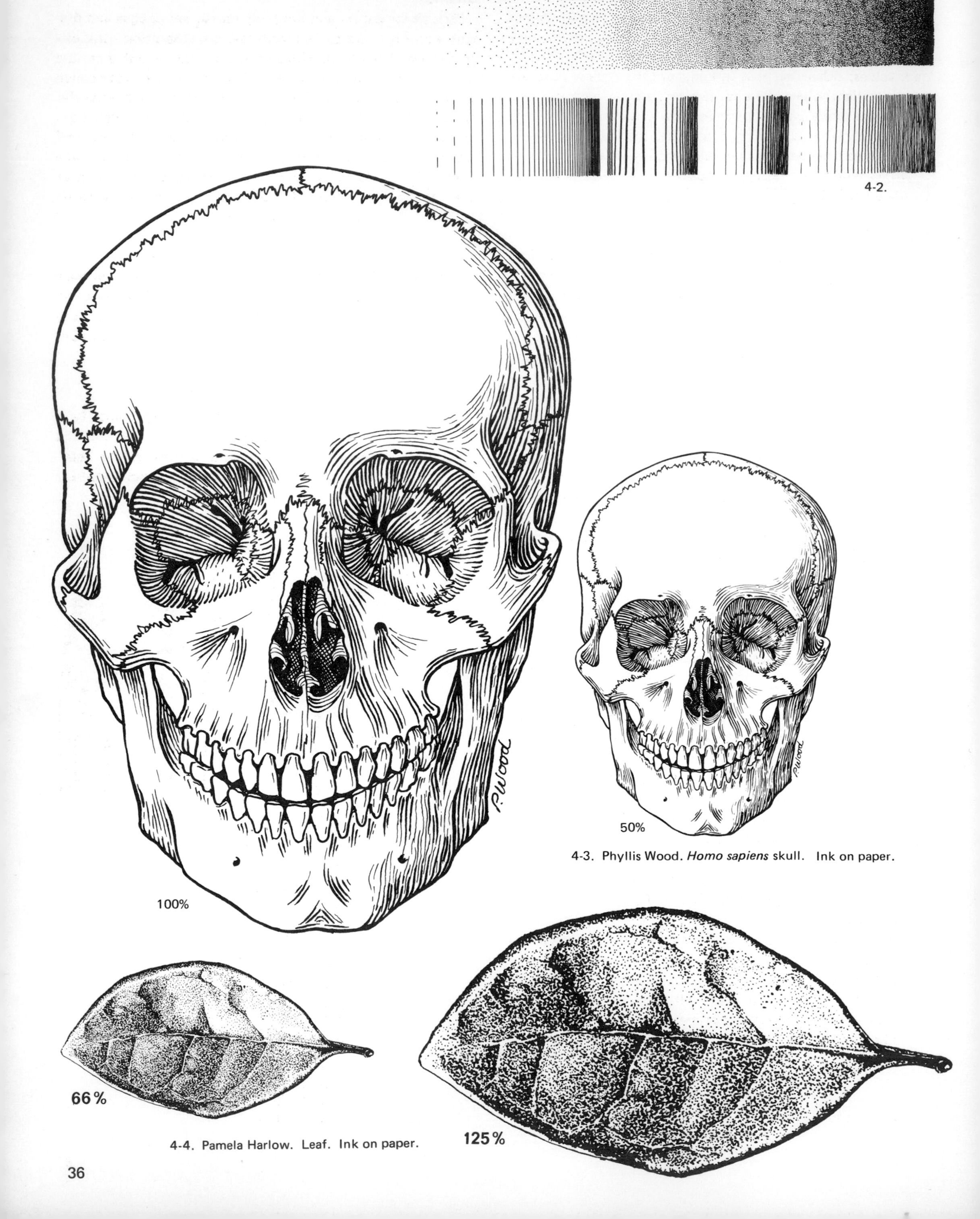

4-2.

4-3. Phyllis Wood. *Homo sapiens* skull. Ink on paper.

4-4. Pamela Harlow. Leaf. Ink on paper.

VALUE

The purpose of black lines and stipples is to give the viewer an impression of various values of gray. Using either heavy lines and dots or lines and dots that are close together creates dark values; either fine lines and dots or lines and dots that are far apart, light values. The varying amount of black ink relative to the white paper determines the position on the value scale (4-2). When a drawing is reduced, the lines and dots in shaded areas merge together and appear as areas of gray. Drawings with line or stipple shading are in effect "screened" by the artist (4-3).

REDUCTION

Ink line and stipple are usually best rendered one and a half to two times larger than the printed size. The reduction process improves the drawing by diminishing imperfections and by crisping up the lines and dots, making them appear as various values of gray (4-4). Enlarging the drawing has just the opposite effect. Imperfections are emphasized, and what appears to be a perfectly good line in the original becomes blotchy and ragged.

When the drawing is reduced, all the ink lines and stipples are closer together. If they are rendered very close together, they may tend to "close up," or to print as uneven blotches instead of as the intended gray values. Each line or stipple must be drawn carefully and precisely in a spatial relation to every other line or stipple. Each stroke or dot must be positioned purposefully, avoiding sketchiness. The blacks must be clear and intense, not weakened by uncertain lines or diluted ink. The white ground must be pure white, not tending toward yellow or smudged in any way. The width of the line is reduced along with the size of the drawing, and, if the lines or stipples are very fine, indecisive, or gray, they may fade out or become lost in the printing process.

MATERIALS

Following are some specific suggestions for your materials.

1. Ground: plate-finish Strathmore; plate-finish Bristol; hot-press illustration board; two-ply, hard-finish tracing paper; Cronaflex, Herculon, Bruning Sure-Scale frosted film; scratchboard; video board
2. Flexible pen: Gillott's 290, 291; Hunt 104; crow quill
3. Mechanical pen: Castell TG 00, 0, 1, 2; Koh-i-noor Rapidograph 00, 0, 1, 2
4. Brushes: Winsor & Newton series 7, sizes 00, 0, 1; Grumbacher series 197, sizes 00, 0, 1; Finepoint series 9, sizes 00, 0, 1
5. Ink: Pelikan T for flexible pens, inks especially recommended for mechanical pens
6. Scraper: Bard-Parker surgical scalpel handle #3 with blade #15, X-acto knife with blade #16
7. Erasers: Art gum, electric eraser with white pencil-eraser fillers

Grounds

Plate-finish papers and hot-press boards are opaque and present a sturdy, hard surface with low ink absorption. Ink can be removed from them with an electric eraser or with a limited amount of scraping with a blade. You can transfer your pencil drawing onto them by either the single- or the double-transfer method (2-40, 2-41) or by tracing papers with a light box. Translucent grounds, vellum, and frosted films allow the pencil drawing to be traced directly. They should be mounted on a heavier board for protection. Clay-coated grounds such as scratchboard, scraperboard and video board allow ink to be scraped very precisely from the surface.

Pens

Flexible pens vary in their flexibility and therefore in their ability to produce lines of varying widths (4-5). The Hunt 104 is the most flexible, followed by Gillott's 290 and 291; the crow quill is the stiffest. Mechanical pens produce lines of unvarying widths in many sizes.

Brushes

Red sable watercolor brushes are full and fine-pointed (4-6). They should be tested with water for their pointing qualities before being purchased. A size 0 brush is sufficient for starting. The same series (shape) of brushes is recommended for wash drawing.

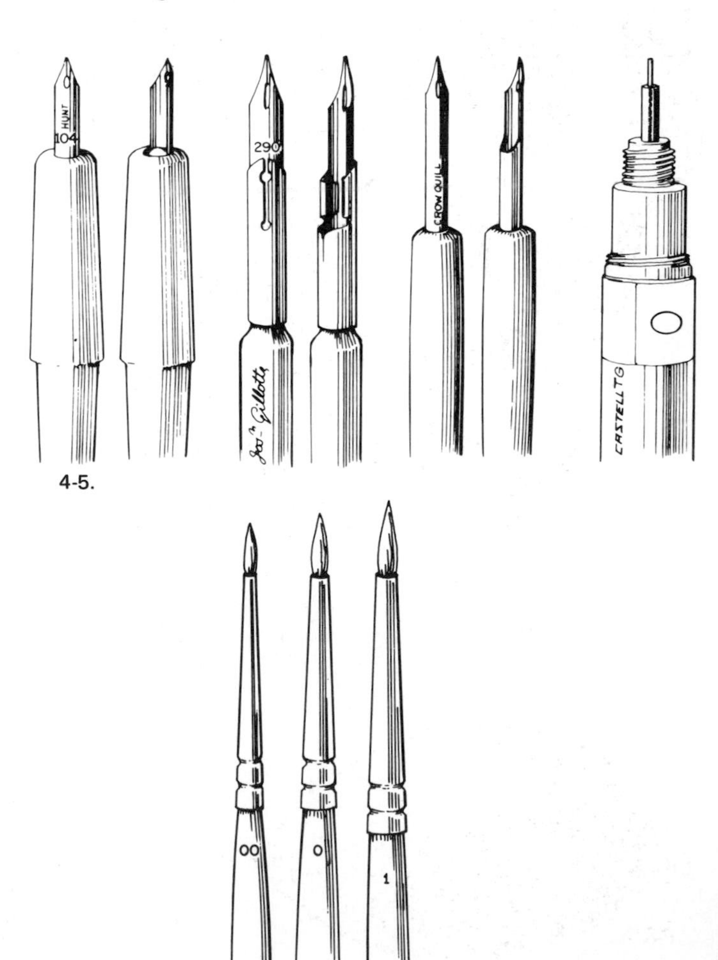

4-5.

4-6.

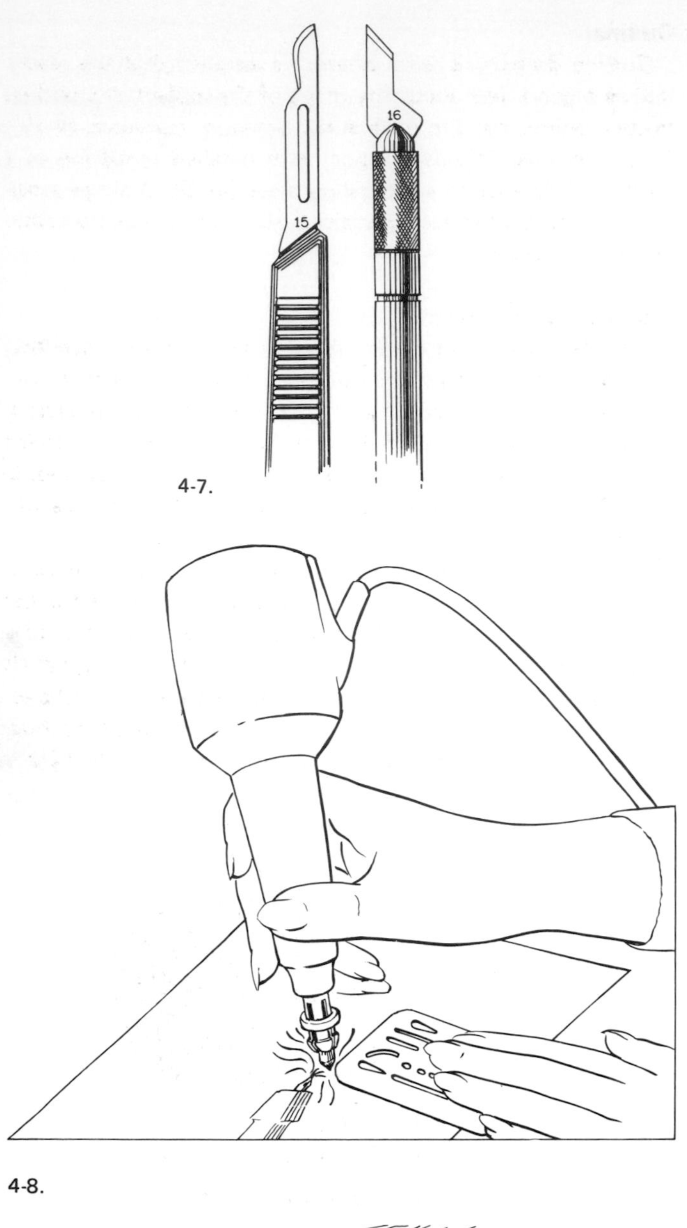

4-7.

4-8.

4-9. B. John Melloni. Eye. Ink on scratchboard.

Ink

Pelikan T ink is recommended, as it is very dense and opaque and usually produces a black line with a single stroke.

Scrapers

Scrapers are used to remove small areas of ink. The blades should be sharpened or replaced as soon as they become slightly dull (4-7).

Erasers

The art-gum eraser is used to clean the ground and erase pencil marks without abrading the surface. The electric eraser is used to erase ink marks. White pencil-eraser fillers are just the right consistency to remove completely dried ink without marring the ground. Ink-eraser fillers are too coarse and dig into the surface. Use the electric eraser carefully, for, if it is in contact with the paper for too long a period, it will heat up and burn it. It is also possible to erase a hole right through the paper. An eraser shield is a handy implement that allows you to erase cleanly up to the point that the shield is protecting (4-8). By handling the electric eraser delicately and using a good-quality paper you can ink and erase several times in the same area without marring the surface.

LINE AND STIPPLE

A drawing can be done in simple outline, shaded with brush line, pen line, crosshatching, or stipple, or executed in any combination of these techniques. By studying the specimen and considering the audience, publication policy, and limitations you should be able to select the most appropriate technique or combination of techniques. Experience combined with logical thinking before touching ink to paper can prevent many false starts.

Line shading

Study the surface of the subject. Are there natural contours or growth lines that would relate to line shading? When using line shading, the direction and character of the lines should not be arbitrary but should help to describe the shape (4-9). Some subjects have a natural contour that you are obliged to follow; others are a matter of choice or of trial and error. Some surfaces such as fur are obviously composed of "lines."

Stipple shading

If the surface is smooth and not easily described with line shading, you should probably render it in stipple (4-10). Stipple can show smooth or textured surfaces and is also useful for depicting a surface coloration or pattern. Only in rare instances would I recommend stippling an entire drawing, since it is tedious and time-consuming. It is often used very effectively in combination with line to show an underlying structure, a basic structure, or a structure with a different texture (4-11).

4-10. Sharon Feder. Ink on paper.

Outline

Outline drawing is used when it is assumed that the reader knows a good deal about the shape of the subject: the outline merely points out the differences between specimens (4-12). It is also used in combination with detailed rendering as a point of reference to surrounding areas (4-13). A single sensitive outline can express a great deal about the shape, bulk, and texture of a subject.

Interpreting the subject

After becoming acquainted with the various ways of handling ink you must learn where, how, and how much of it to use. The tendency is to become so delighted with the rendering process that one embellishes on and on. One needs restraint and a knowledge of when to stop. The ink line or stipple should capture the essence of the subject without necessarily including every detail.

You must know not only your subject but also your audience. Don't draw the whole elephant if all you need is the trunk. The focal point of the drawing should be more fully rendered, with more detail than in the peripheral parts, which serve to relate rather than to inform the viewer. It can be rewarding in terms of both design and information to combine a simple outline with tightly detailed line and stipple (4-14).

4-11. Carol Angell. Fir cone. Ink on paper.

4-13. Jane Rady. Oyster shell. Ink on paper.

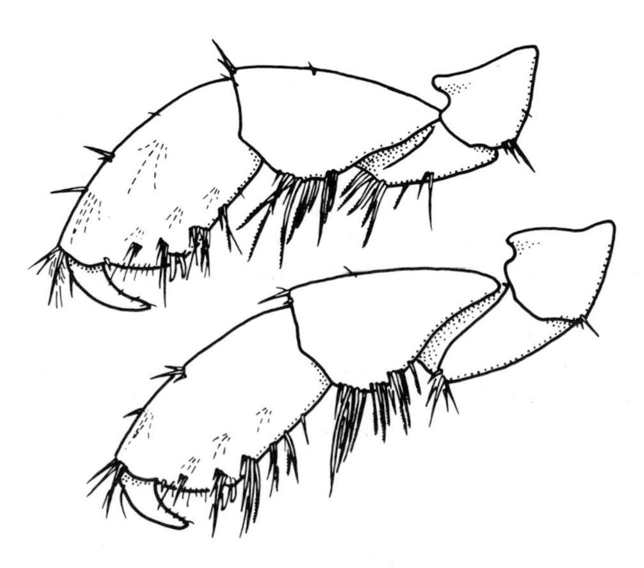

4-12. Craig Staude. Freshwater amphipod legs. Ink on paper.

4-14. Patricia Veno. Ink on scratchboard.

TECHNIQUES

Settle yourself into the place that you have carefully prepared for executing your drawings. You must have clean, steady hands (soap and water but no coffee); your work space should have: (1) a clean, smooth, uncluttered, slightly tilted drawing board, (2) a good filtered light at your left (right for left-handed people), and (3) a flat surface close by for ink, water jug, and tissues (4-15).

Reproduce the following exercises in the same size as the originals. Practice tracing them line for line. Tracing other artists' work is very helpful in training your "hand." There is no danger of copying another's style, as it is similar to handwriting: after you learn the basics, you will develop your own signature or personal technique.

4-15.

Pen line

While pen and brush can often be used interchangeably, short, weighted (varied widths) and unweighted (same width) lines, particularly those with tight curves, are best done with a pen because of its flexibility and its ability to turn smooth or sharp corners (4-16).

Dip only about 75% of the pen into the ink and withdraw it, touching the clean neck of the bottle to release some of the ink (4-17). With this amount of ink in the pen you have more control, can make finer lines, and can avoid ink blobs on the paper. The pen should be rinsed in the water jug and dried on a soft tissue or cloth every few minutes in order to retain a crisp line. India ink dries very quickly and, if left to dry in the pen, it produces an uneven or blotchy line. Do not soak the pen in water, however, as this will ruin the holder.

Hold the pen easily, not tensely, at a very low angle to the paper. Use the little finger as a "slider," helping to hold the pen at the right distance from the paper. It may be helpful to place a small piece of paper under the heel of your hand for a smoother motion. Both nibs of the pen should contact the paper with equal pressure.

Strokes are made in a natural arc away from the body. Long strokes are made by moving the arm in an arc from the elbow (4-18). Medium-sized strokes are made by moving the hand with the heel as the fulcrum (4-19). Smaller strokes are made by using only the fingers to complete the arc (4-20).

To make an arc in the opposite direction, turn the paper so that your hand and arm always move naturally: the artist stays in one place; the paper is moved. It is important to keep the paper free, not taped down, so that it may be moved to facilitate the easiest, most natural strokes.

When joining two lines or continuing a natural contour that has been interrupted, start the new line inside the end of the previous line in order to prevent a jerky appearance. Practice making parallel lines varying from close together to far apart. Avoid hooking the end of the stroke (4-21).

4-16.

4-17.

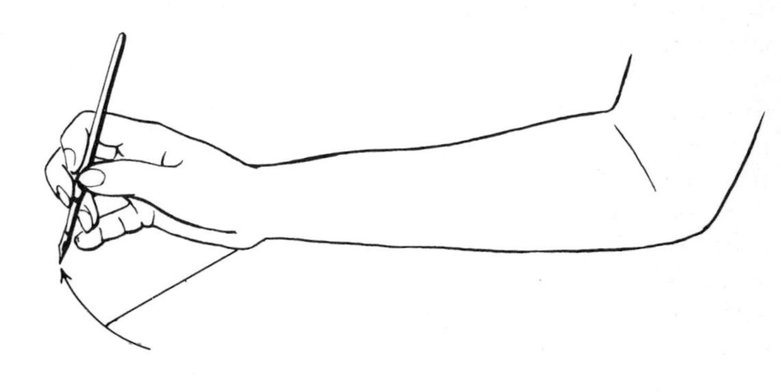

4-19.

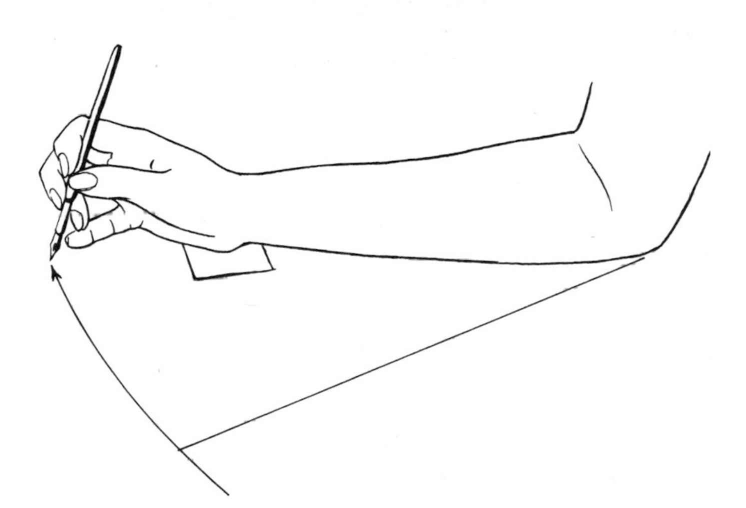

4-18.

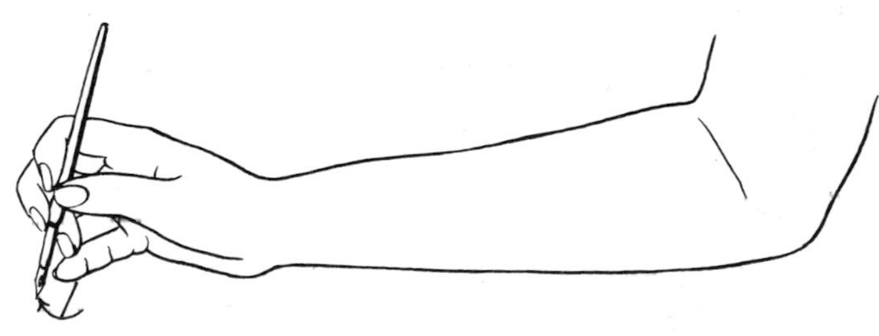

4-20.

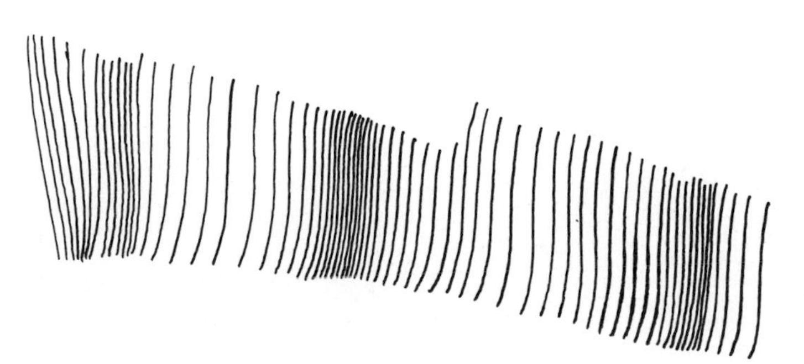

4-21.

Weighted line

By letting the pen glide over the paper with no pressure other than the weight of the pen you can draw a very fine line. Varying pressures produce varying lines from very fine to very wide. The greatest variation is possible with more flexible pens. Weighted-line drawings can have much vitality and character. Practice many variations. Each line may be made in several steps, doing the thin parts first and then adding the heavier part (4-22).

Unweighted line

An unweighted line is made by exerting an even pressure on a flexible pen or with a mechanical pen. A mechanical pen (sometimes referred to as "the pen without a soul") produces a perfectly constant weight of line, with the width dependent on the pen size. Pens vary from size 000 (very fine) to size 5. Sizes 00 to 2 are the usual range used for drawing. Mechanical pens must be kept scrupulously clean and filled with the recommended brand of ink. If the pen becomes clogged, it can be disassembled and washed thoroughly in soapy water, dried, and reassembled.

Unweighted line offers only one option for variety: the number of lines and their distance from each other. More lines instead of heavier ones are therefore required in the shadow areas. Care must be taken not to give a mechanical appearance to the drawing (4-23).

Crosshatching

In crosshatching opposing directions of lines are at oblique angles to one another. Right-angled crosshatching produces a wire-screen effect and should be avoided. Two, three, or four directions of lines may be used as the shadow becomes darker. Dots may be placed in the centers of the little parallelograms (4-24).

4-24. Deanna Manley. Barnacles on rock. Ink on paper.

4-22.

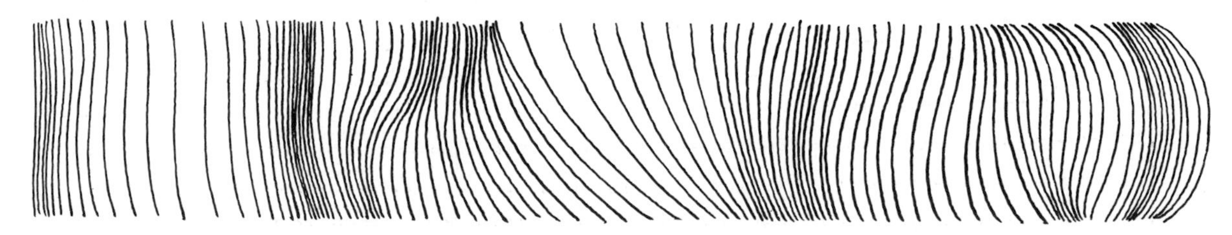

4-23.

Brush line

The brush is handled in much the same way as the flexible pen, but it especially shines when long, smooth, or weighted lines are needed. The ink flow can be precisely controlled to make very heavy to very fine lines. Any slight variation in the way in which the brush is held produces a change in the weight or character of the line (4-25). Practice drawing long, smooth, varied lines (4-26).

Dip the brush only part way into the ink and release some of the ink on the neck of the bottle. Hold the brush more vertically than the pen for more delicate control—at almost a 90° angle to the paper.

Rinse the brush in the water jug and dry it often on a soft cloth or tissue as you work. Do not leave a brush with the bristles resting in a water jug, as they will soon become permanently bent or separated: the brush will never again give a controlled line and must be demoted to glue dispenser. Brushes should be left to air-dry in a pointed position after being washed in mild soap and water. If they must be carried from one place to another, an easy way to protect the bristles is to tie them together, using the handles as crutches (4-27). It is necessary to invest in good brushes for this kind of precision work. They will last for years if treated with respect.

4-25.

4-26.

4-27.

Line-shaded drawing

It is a good idea to pencil in the general direction of the shading lines on your preliminary sketch, but it is not necessary to trace each line laboriously in pencil before doing your ink drawing (4-28). With practice you will become adept at determining the width of the lines and the distance between them. Heavier lines seem closer to, thinner lines farther from the eye. Lines that are closer together recede from the eye; farther apart, advance toward it.

A good exercise for determining the direction in which to shade a subject is to start with a continuous-tone drawing and to line-shade it in several different ways (4-29). You can then choose the rendering that seems most representative of the character of the subject. If part of the line shading is difficult to interpret, place a piece of tracing paper over the trouble spot and experiment with several directions until one seems right.

In addition to the *direction* the *character* of the shading lines should be meaningful. The personality of the line can show roughness, smoothness, delicacy, color, or coarseness (4-30).

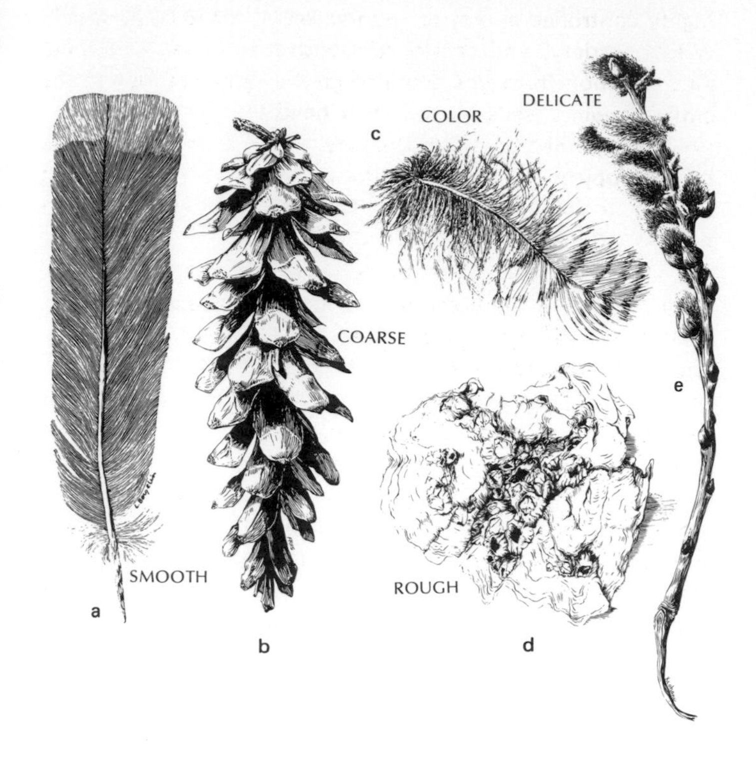

4-30. a. Lynn Bergelin. Eagle feather. Ink on scratchboard. b. Anton Friis. Cone. Ink on scratchboard. c. Pamela Harlow. Feather. Ink on paper. d. Elizabeth Keohane. Oyster shell with barnacles. Ink on paper. e. Elizabeth Keohane. Pussy willow. Ink on scratchboard.

4-28. Paula Richards. Garlic. Ink on paper.

4-29. Beverly Witte. Garlic. Ink on paper.

Controlled and casual treatments

The choice between a tightly controlled and a looser, more casual treatment must be made with every technique, but the difference is most apparent with ink-line shading (4-31). A highly controlled approach requires every line to be perfect in weight, variety, and spatial relationship to every other line. Any variation from this precision gives an amateur look to the entire drawing, while on the other hand there is the danger of producing a mechanical appearance, which is undesirable in a natural subject. It is very helpful to practice control, but it is difficult to achieve perfection. Each line must stand alone in meaningful simplicity.

A casual approach is at the opposite end of the style spectrum (4-32). While every line is in its own area and relates perfectly to every other line, there is no rigidity to the drawing. Any artist who draws in this style interprets the subject differently: it has a certain light-hearted expressiveness that is particularly charming. The seeming frivolity of line does not mean, however, that this style is any less difficult to achieve. There is a danger of producing a sketchy drawing, inappropriate for scientific illustration. Each line has importance, but additions and subtractions are left to the discretion of the artist.

Most pen-and-ink drawings fall somewhere between the extremes of tight and loose rendering.

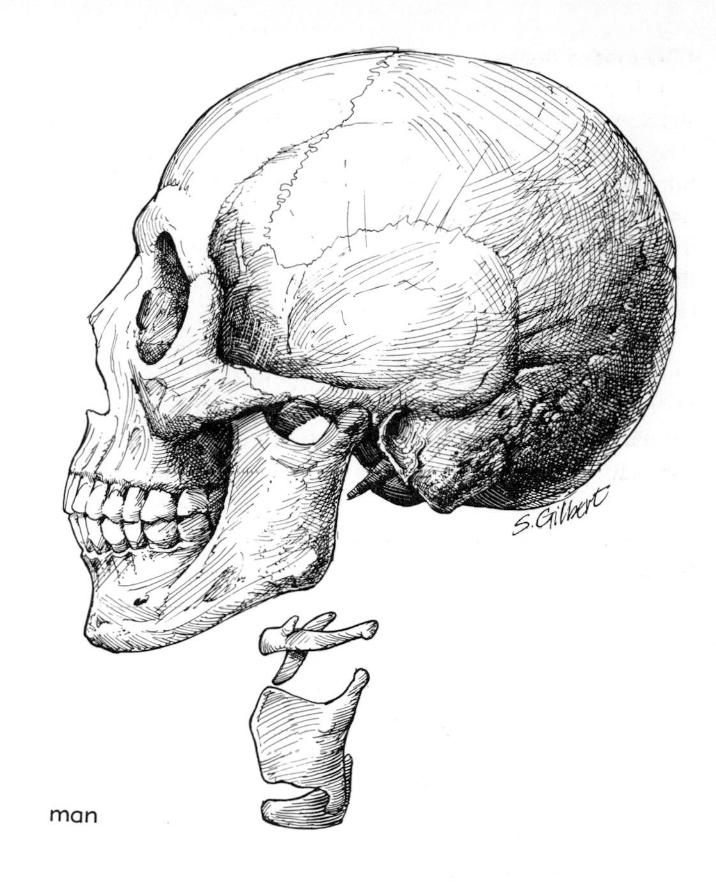

4-32. Steve Gilbert. Skull. Ink on paper.

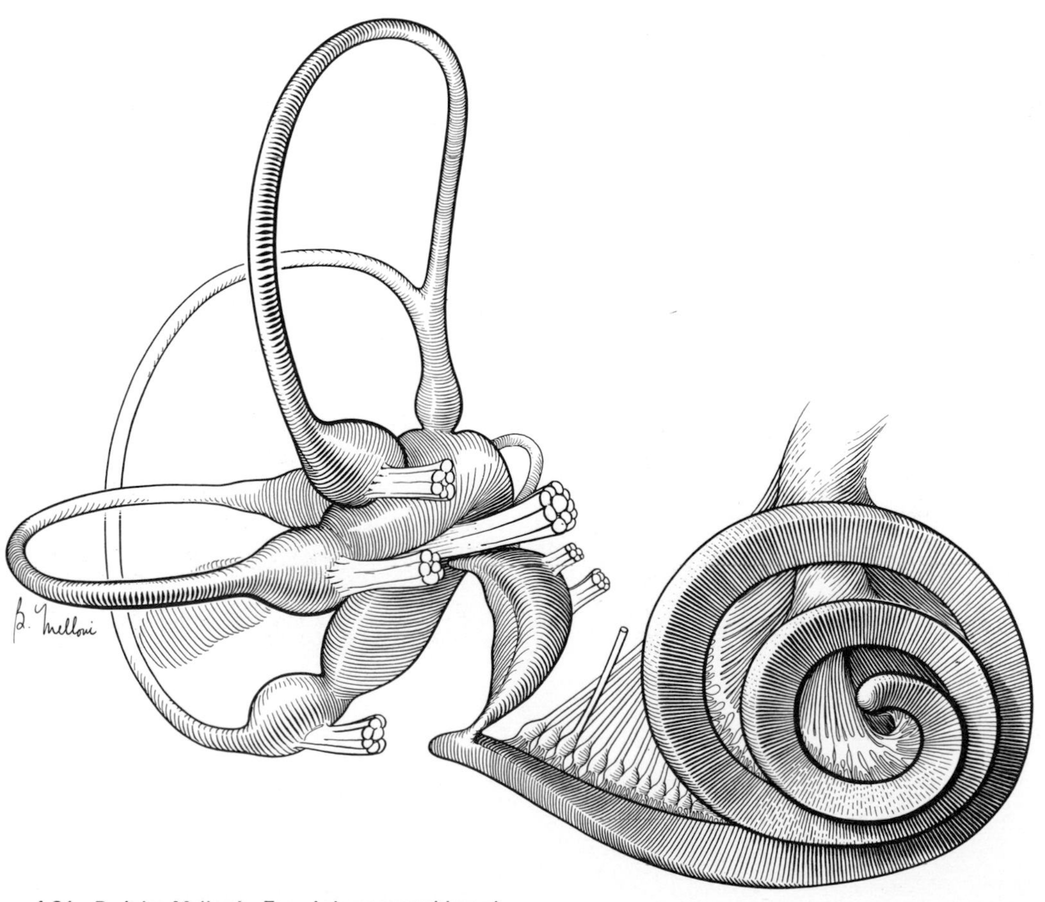

4-31. B. John Melloni. Ear. Ink on scratchboard.

4-33. Deborah Moffit. Ink on paper.

4-34. Deborah Moffit. Ink on paper.

Stippling

Stippling is done with a stiff pen such as a crow quill or with a mechanical pen. The mechanical pen renders all the dots the same perfect size and shape but does not easily allow for a change in the size of the dot (without changing pens). Beginners usually prefer stippling to line shading even though it is much slower, because one does not get into trouble with it nearly so easily. Friends and relatives also marvel over all those little stipples!

To learn the stipple technique, practice drawing a smooth, contoured surface (4-33). The dots are carefully placed equidistant to each other and become increasingly closer until they merge. Every stipple is placed in a precise spatial relationship to every other stipple. Stipples do not touch each other except by design. They are seemingly placed at random and not in any mechanical order.

Stipples are placed in lines only when this arrangement is meaningful to the shape (4-34). A stippled drawing may be stippled around the outline or have a line around the edge (4-35). Rough surfaces are stippled with uneven spacing or with different-sized dots (4-36). Various values of surface coloration may be shown by continuing the contour relationships of the stipples and at the same time abruptly changing the specific percentage of stipples (4-37). Stippling can be an effective technique for rendering only a specific part of an outline drawing (4-38). Phantoming structures below the surface and showing a part of the drawing in a different surface texture are valid uses of stippling (4-39).

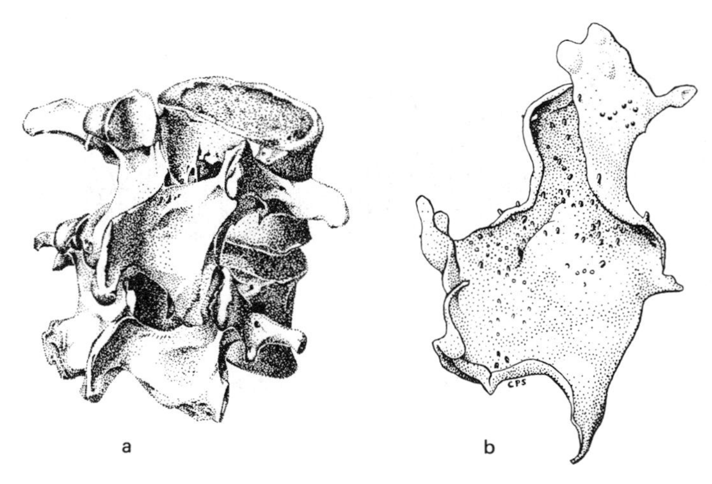

4-35. a. Margaret Watson. Vertebrae. Ink on paper. b. Craig Staude. Seaweed. Ink on paper.

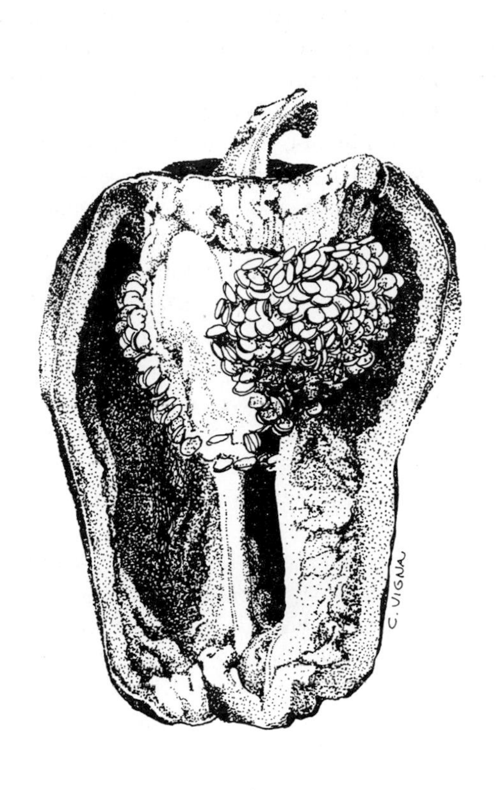

4-36. Cheryl Vigna. Green pepper. Ink on paper.

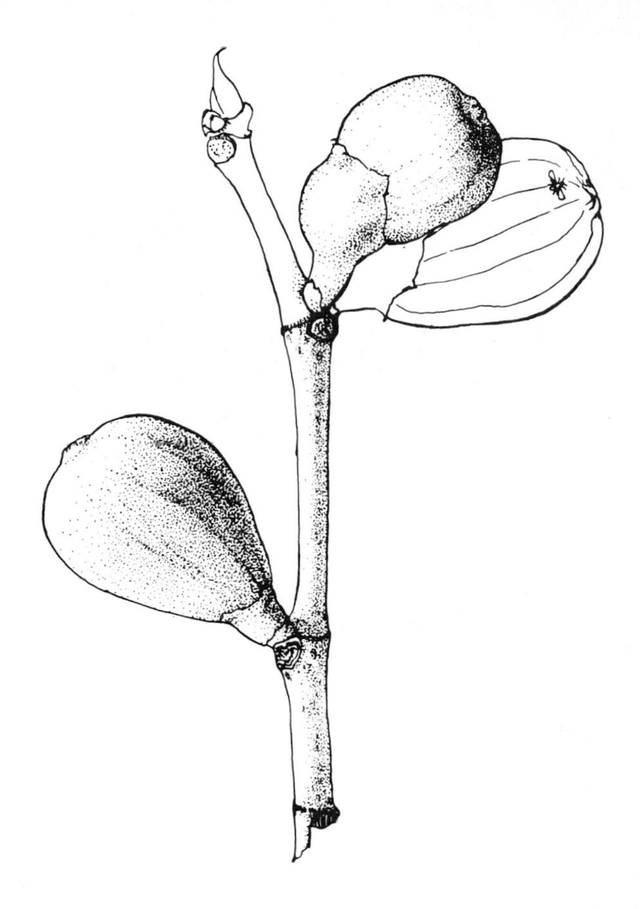

4-38. Kathleen Skeele. Fig. Ink on paper.

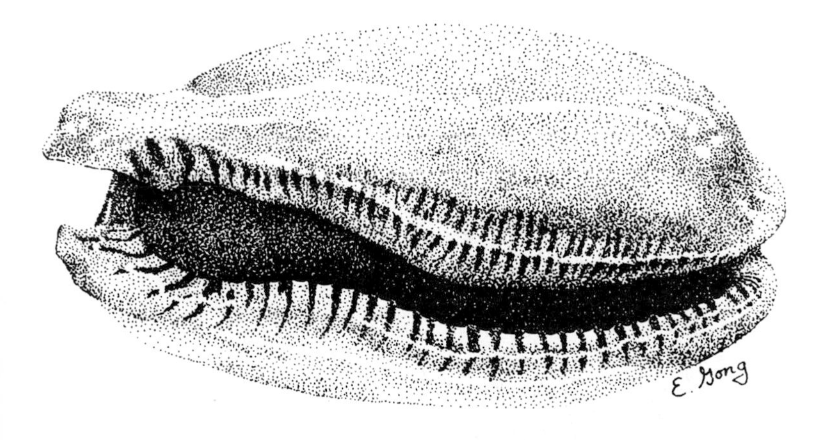

4-37. Elizabeth Gong. Shell. Ink on paper.

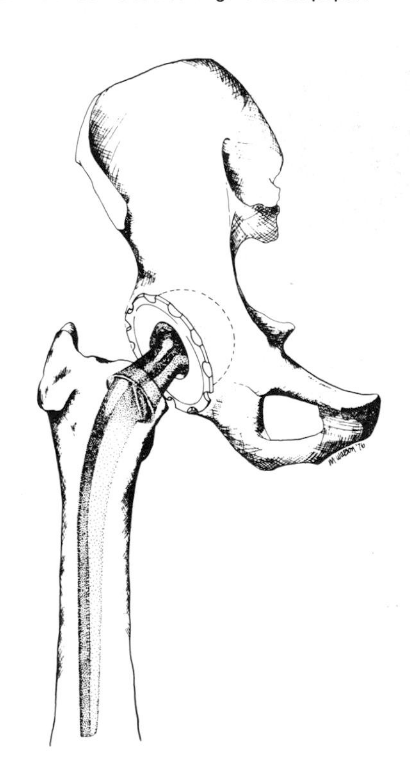

4-39. Margaret Watson. Hip-joint replacement. Ink on paper.

Depicting distance

In studying atmospheric perspective we learn that, as an object recedes from the eye, it becomes less clear, dimmer, softer, more hazy. While this principle would seem to apply only to greater distances, it can also be used effectively to show very small differences in distance in a small specimen. The part of the specimen that is further away from the eye is drawn with finer lines than is the part closer to the eye. This convention gives a three-dimensional quality to the subject. The further away an element is from the viewer, the lighter the rendering should be (4-40).

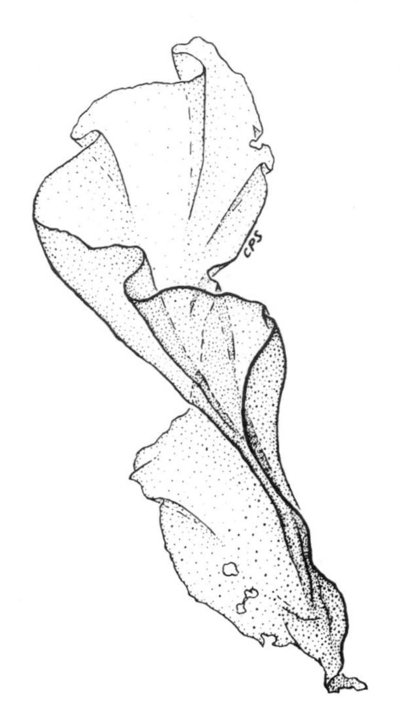

4-40. Craig Staude. Seaweed. Ink on paper.

4-41. Phyllis Wood. Dollar plant. Ink on paper.

A useful convention for showing which part of a solid or translucent subject is closer to the viewer is to break the most distant line. When one line is broken, it becomes apparent that the unbroken structure is closer to the eye (4-41). The missing part of the line can be left out or removed carefully with a scraper. When using the scraper, take care to disturb only the unwanted ink and not to remove too much of the paper surface. Such a removal is irrevocable: scraped paper cannot be drawn on again. Scraping does leave a crisper edge to a broken line than does stopping the line. Obliterating a line with white paint is generally too clumsy.

Detail

Almost any subject can be rendered in a three-dimensional way. For example, leaves have varying thicknesses; even a thread has more than two dimensions. It is often very effective to show that thickness with parallel lines, one on the shadow side and a thinner one on the lit side. Just a suggestion of one of the lines is sometimes sufficient to show the three-dimensional quality.

When drawing single fine plant or animal hairs, the ends may almost disappear. In order that they may be left on the printed page, I often give the end of the hair extra width. While this may offend the purist, the only alternative is often to omit the full extent of the hair, which may be an essential characteristic (see 4-45).

Scratchboard technique

Drawing with ink on scratchboard is a different and marvelous experience. Lines can be produced in two ways: by putting a positive ink line on the board and by removing the ink in a negative line from the board (4-42). An area can run from black to black with white lines to white with black lines to white.

While it is possible to scrape away parts of lines when using other grounds, this technique is most effective with scratchboard. On scratchboard it is also possible to produce crisp white lines from an inked area. Scratchboard has a fragile clay-coated surface that is the basis for its versatility.

Procedure

Mount the scratchboard on a piece of sturdy cardboard, as it cracks easily if bent. It is a good idea to dry both the scratchboard and the mounting board in a dry-mount press to avoid subsequent curling. Clean the scratchboard thoroughly with an art-gum or Pink Pearl eraser. Do not touch the board with your fingers, as this will affect the ink line.

The pencil composition is transferred to the scratchboard by the double-transfer method (2-40). This is done with as few lines and as lightly as possible, so as not to form grooves or otherwise mar the delicate surface of the board. After transferring you can carefully strengthen the pencil lines that you want to keep and remove any marks that are not necessary.

Put in the shadow areas in solid black, using a brush and making sure that the ink is applied smoothly with no rough places or gobs of ink. Any variation in the smooth application adversely affects the character of the scratched line (4-43).

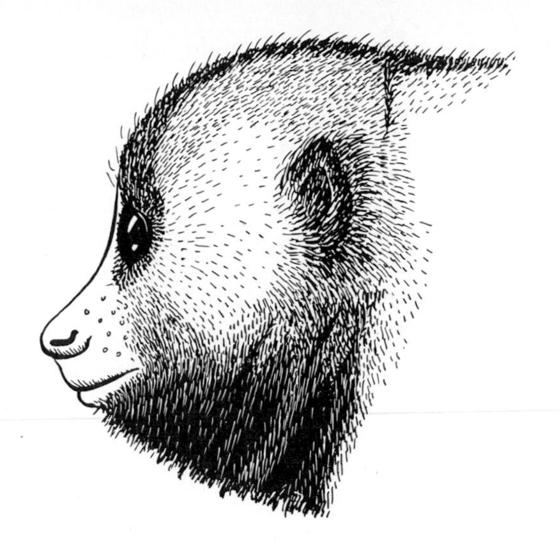

4-42. Phyllis Wood. *Nycticebus coucang* (slow loris). Ink on scratchboard.

4-43. Solid black areas rendered.

After all the shadow areas are blackened in, the black lines are rendered (4-44). The white scratch lines are then produced in the shadow areas by carefully removing the ink with a #16 X-acto knife or a Bard Parker #15 surgical blade (4-45). Do not remove more of the starchy surface than is necessary and do not gouge the surface. If this step is done with a gentle hand, the area may be reinked and rescratched several times if necessary. For the cleanest, crispest lines, however, it is better to do all the inking before removing any of the ink.

It is best to use a brush on scratchboard, because it will not harm the surface. If you do choose a pen, take care not to scrape the scratchboard. The pen does pick up a certain amount of the surface, however, and must frequently be rinsed in water and dried as you work. To avoid leaving finger and hand marks on the board, keep a slip of paper under your hand as you work.

Scratchboard is the preferred technique for white lines. It is particularly suitable for rendering fur, as the hair pattern can be shown in both the light and the dark areas. Very smooth, thin lines can be produced, and unwanted ink and smudges can be removed easily.

Other grounds

The scratchboard technique can be used on other materials. Frosted films such as Cronaflex, Herculon, or Bruning Sure-Scale drafting film and clay surfaces such as Video or Media papers can be used. These are not rescratchable, as the surface is very thin, but they are many times less expensive than any brand of scratchboard and may serve your purpose just as well. The films have an added advantage, as you can trace your original drawing directly, omitting the transfer step. Clay-surface boards and papers should be mounted before using. Opaque boards should be dry-mounted or rubber-cemented on white or colored board. Films need the stability and reflective qualities of a white mounting board.

4-44. Black lines added.

4-45. White lines scratched. Phyllis Wood. *Ateles geoffroyi* (spider monkey). Ink on scratchboard.

Coquille-board technique

To simulate halftones, a very quick and effective but seldom used technique is to use coquille board and a litho crayon (4-46). By drawing the waxy black litho crayon over the pebbled surface of the coquille board black dots similar to ink stippling are produced. This technique is many times faster than hand stippling, but it takes some practice to keep the drawing crisp and unsmudged.

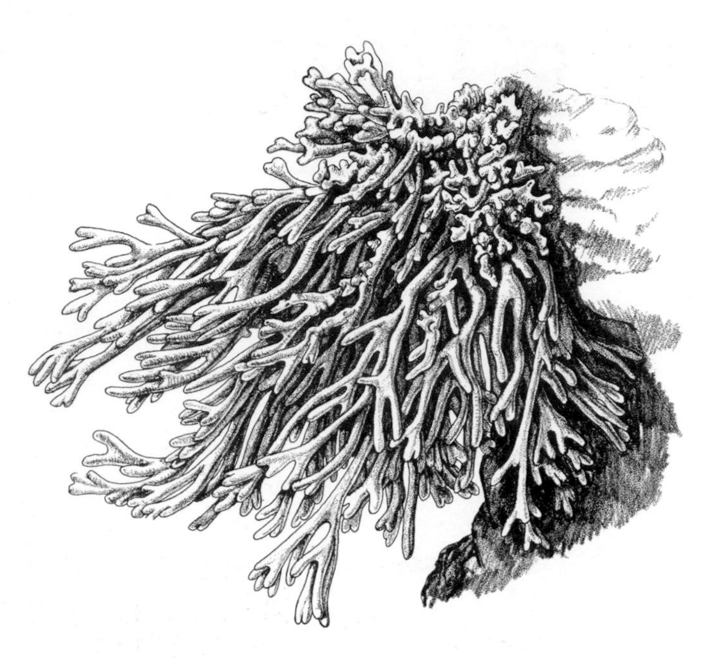

4-46. Janet MacKenzie. ***Codium***. Litho crayon on coquille board.

Materials

Following are some recommendations for coquille-board materials.

1. Ground: coquille board #1 (fine-textured)
2. Crayon: William Korn's litho crayon #5 (copal, hard)
3. Transfer paper: Saral
4. Designer's gouache, Permanent white
5. Mechanical pen: #000

Coquille board comes in coarser grades and in square- and oval-shaped patterns, but the fine grade has a consistent dot pattern that is best for scientific drawing. The #5 dense black, paper-wrapped litho crayon is used, as it does not smudge as easily as the softer (#1, #2, #3, #4) crayons.

Procedure

Transfer the preliminary drawing to the coquille board with a 4H pencil, using nonindelible Saral transfer paper. Press very lightly and lighten the marks with a kneaded eraser if they are too dark.

Render any black outlines or shadows with the #000 pen. It produces a crisper edge than is possible with the litho crayon alone.

It is important that the litho crayon be kept sharp at all times. Sharpen it first with a razor blade or scalpel and then on a sandpaper block. To render the gray values, draw the crayon over the board at low angle, grazing the pebbly surface. Rolling the crayon as you work helps to keep the point sharp (4-47).

Apply the various values by gradually working them up, as they cannot be lightened. Erasures are not possible with this technique: they will only smear the waxy crayon. Designer's gouache, Permanent white is very dense and can be used to obliterate any unwanted blacks.

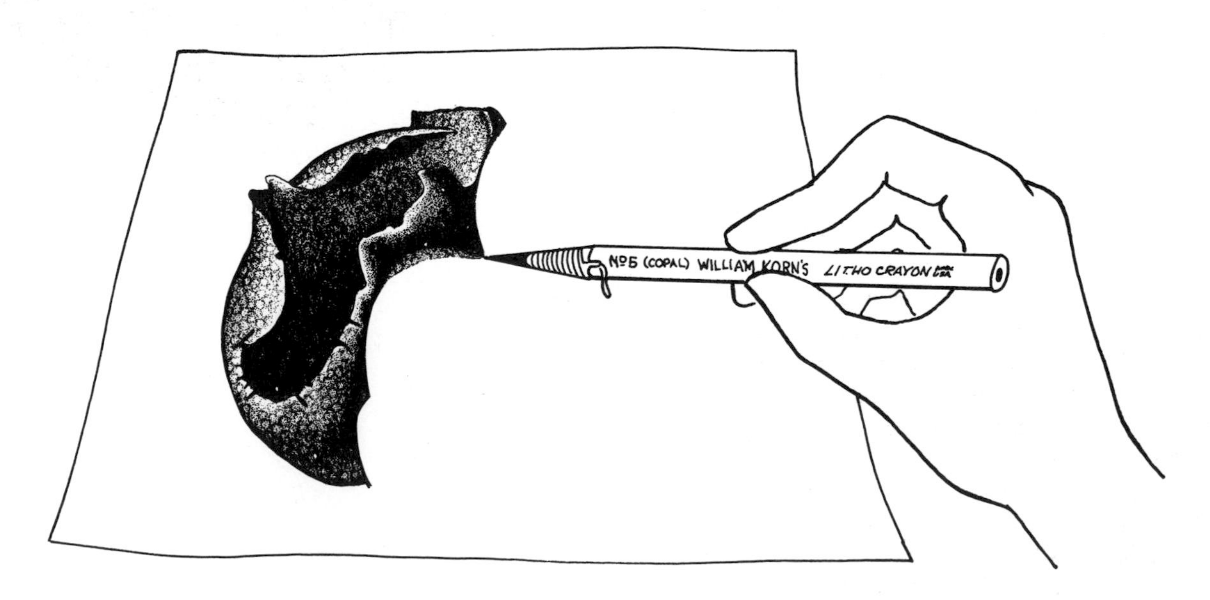

4-47. Janet MacKenzie and Phyllis Wood. Ink on paper and litho crayon on coquille board.

CHAPTER 5.

Continuous Tone

Continuous tone is a term applied to rendering techniques that utilize varying values of gray, including wash (watercolor with only black pigment), carbon dust, pencil, and airbrush. By using continuous tone the artist can realistically interpret the actual shape, color values, surface texture, and character of a subject (5-1).

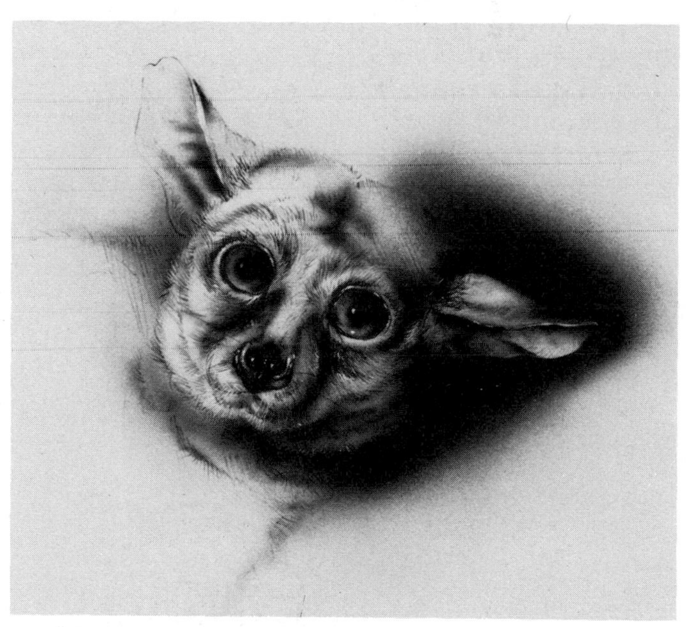

5-1. Joel Ito. *Galago crassicaudatus* (bushbaby). Airbrush and watercolor on film.

REPRODUCTION

To be reproduced, a continuous-tone drawing must be screened in the same manner as for halftone photography. In halftone photography gray areas are broken into black dots (in a screen pattern): small dots for light values, medium dots for middle values, and large dots for dark values.

Producing a good screened plate is a technically demanding job. For this reason the artist cannot always expect the print to be an exact replica of his original rendering. The reproduction camera is very critical: it detects and photographs all detail including corrections if not handled properly and also "sees" the artwork somewhat differently than your eye may see it. Book illustrations are usually quite faithful to the original, but the same cannot be said for all scientific publications. Subtle shading, delicate detail, and pale values may be lost in the reproduction, so the artist must be bold rather than timid in his rendering. The original drawing should be planned to allow for little reduction: between 90% and 75% of the original is recommended.

Being aware of the hazards of continuous tone, anticipating the delight of having a perfectly reproduced continuous-tone drawing, and realizing that in many instances a specimen can be interpreted only by using one of the continuous-tone techniques, it is important to understand and to master them.

WASH DRAWING

Wash drawing can be a fast and effective way to render a specimen. It is particularly applicable to depicting smooth gradations and even tones while at the same time interpreting fine detail. I feel that the use of water as the vehicle for the pigment makes wash a natural technique for drawing living tissue—botanicals and zoological forms (5-2)—while dry techniques such as carbon dust and pencil are more appropriate to dried botanicals, bones, and shells.

Elements

A wash is a variable combination of three elements: (1) the ground is the board or paper that you are working on: it provides the light or luminosity in your rendering; (2) water is the vehicle that carries the pigment and provides the wet or sparkly look characteristic of a wash drawing; (3) pigment defines the shape by creating the dark values and shadows. You must control the paint-loaded water by persuading each particle of pigment to rest on the ground where you want it. The particles are floated on, never scrubbed into the ground. Think of wash as a living thing and work quickly to control it before it dries. If worked quickly and confidently and not overworked, it will maintain its characteristic fresh and spontaneous quality.

Using the techniques and materials described here, you can work up your drawing gradually. You can go back, strengthening values, altering relationships, and adding details. A finished wash drawing is relatively permanent and does not mar easily if treated with care.

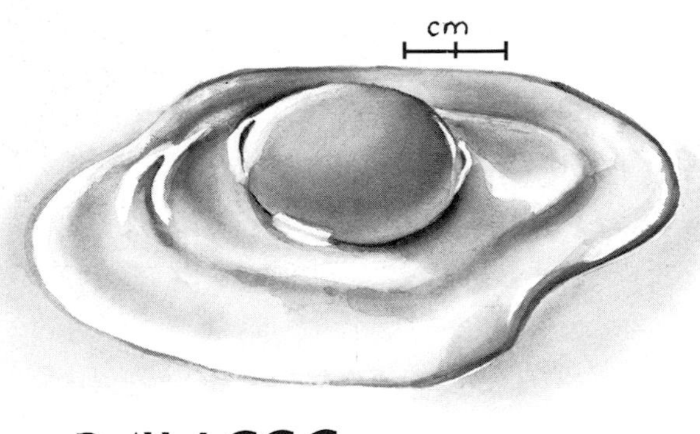

5-2. Kathleen Todd. Wash on cold-press illustration board.

Materials

Following are some recommendations for wash-drawing materials.

1. Ground: cold-press illustration board, medium or heavy weight; hot-press illustration board, medium or heavy weight; 100% rag watercolor board
2. Pigment: Weber permo-black lampblack watercolor (tube) or other black watercolor (tube)
3. Brushes: full, pointed, round, 100% red-sable brushes, sizes #0, #3, #5, Winsor and Newton series 7 or 707, Grumbacher series 9, Delta sablon 550
4. Palette: white enamel or china with a large, flat mixing surface
5. Water container: large, at least 24-ounce capacity
6. Soft cloths or tissues
7. Hair dryer: handheld

The ground comes in two types: cold-press board, which has a roughness or tooth, and hot-press board, which is smooth. It is easier to apply a smooth wash to the cold-press board, so start with it. You should later experiment with the smooth hot-press board and the 100% rag, which is wonderful to work on, expensive, and perfect for special drawings.

The Weber pigment is quite permanent and allows you to work back into an area to build up the values without disturbing the underlying values and details.

The proper shape and quality of a sable brush is of utmost importance. Fine sable brushes are expensive but, if treated with respect, can last a lifetime. Test a brush before buying it to see whether it will "point." There should be no "wild" or independent hairs. Never leave the bristles in a bent position or standing in the water container, since they may acquire a permanent wave that can never be corrected. Wash them gently but thoroughly with warm water and soap after each use. Let them dry in a point and protect the bristles between using.

Preparation

Before starting wash your hands thoroughly with soap and water. Any skin warmth or finger moisture transferred to the board will repel the dissolved pigment and make it impossible to obtain a smooth wash. On large renderings you can protect the board from your hands with a paper mask taped to the area that you are not working on. Clean the board by erasing gently with an art-gum eraser, taking care not to abrade the surface. Make sure that the board is free of eraser crumbs, unwanted pencil marks, brush hairs, and pencil dust. Any of these will repel or attract pigment in an undesirable fashion. Make yourself comfortable with a sturdy chair, good upper-left-hand light source (for right-handers), and slightly tilted drawing board. Keep your water container and other materials on a flat table within easy reach (5-3).

Exercises

Every wash drawing is merely a carefully planned combination and refinement of exercises. When you have mastered all of them, you can apply this proficiency to the rendering of your subject.

Smooth wash

The first basic exercise is smooth wash. Dampen an area of the board 2″ square (5 cm square) with a clear-water wash. As you want crisp edges, keep the water within the borders of the square. Allow it to soak in until the board is evenly damp with no puddles or standing water and until the meniscus of the water is below the pebbled surface of the board. To help judge the soaking time, turn the board to the light to see how shiny and wet it is.

When the board is ready for the pigment to be applied, use your largest brush and mix on your palette a sufficient amount of pigment and water to make a medium gray. Don't get any undissolved bits of pigment on the brush, or it will streak. Test the mixture on a scrap of board for the value, remembering that it will be lighter when it dries. Load the brush with the pigment-water mixture and, starting at the top of the dampened area, make a horizontal stroke from left to right (5-4). Guiding the bead of pigment down the side, stroke from right to left, overlapping the first stroke. Repeat this zigzag process until the entire area is covered. You may need to reload your brush with the mixture on your palette. If there is a line or bead of extra pigment at the bottom of the area, pick it up with a semidried brush. If the brush is too dry, you may pick up more paint than you intended and leave a light line at the bottom.

The illustration board acts as a miniature easel. It can be moved and tilted to allow gravity to assist or to slow the flow of pigment. The pigment, however, must flow in only one direction. Any variation in the direction will result in an uneven wash.

5-3.

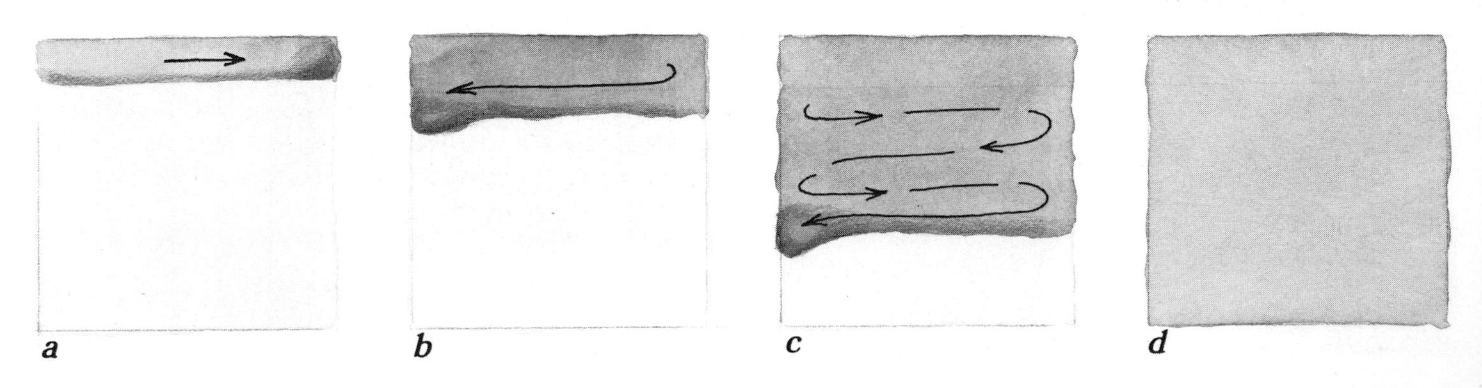

5-4.

5-5.

5-6.

5-7. Smooth values.

5-8. Graded washes.

The clear wash applied before the pigment allows the paint to spread evenly. If it did not soak in long enough, the paint will "cobweb" (5-5). The cobwebby streaks that are so charming in a rendering of a cloudy sky in wet-on-wet watercolor are not desirable in precise scientific illustration. If the clear wash dries for too long, stroke marks will result (5-6). They can be delightfully artistic, but in a drawing of a scientific subject they may be interpreted as part of the form or surface of the specimen. You must confine your artistic license to areas that will not falsely influence the viewer's interpretation of the subject.

After some practice you will be able to judge the right combination of pigment and water for the desired value. If the wash is too light, you can build up the values after the first wash has dried completely by laying in another clear wash in the same manner as for the first one and reapplying the pigment mixture. The value of the wash cannot be lightened as can most watercolors, however, because these pigments are relatively permanent. It is therefore preferable to start out with a lighter rather than a darker wash. Make a value chart of a series of seven smooth washes (5-7).

Graded wash

After you have mastered smooth wash, graded wash is the next challenge. Apply a clear wash and allow it to dry partially in the same manner as for smooth wash. Load the brush with a dark value of pigment-water mixture and make two horizontal strokes. For the third stroke load the brush just as fully but use a lighter pigment-water mixture. Continue this procedure, lightening the mixture very gradually for every other stroke but loading the brush just as fully, down to the lower edge of the area. The tilt of the board, the fullness of the brush, and the very gradual dilution of the mixture combine to produce a smooth, graded wash.

If the area is too light or not entirely smooth, wait until it is completely dry and until all particles of pigment are set and then start again with a clear wash, partial drying, and a second graded wash. A gentle dabbing of a wet area with a tissue will lighten it. The blast of a hair dryer at a crucial point in the drying will control the flow of pigment and "freeze" the changing wash.

Practice this same procedure but start with a light value and go to a dark or with a light to a dark and back to a light (5-8).

Competence in these two basic exercises, smooth and graded washes, is necessary in order to produce an acceptable scientific wash drawing.

Geometric solids

Geometric shapes are the most demanding subjects to render. They are so specific and perfect that any slight deviation is immediately apparent. All sorts of inaccuracies can be overlooked as specimen variations or anomalies in biological structures, but an error in a geometric subject is obvious to all. If you can render geometric solids precisely and freshly, you can render anything.

Draw and pencil-shade a cube accurately on tracing paper. Keep in mind that the light source is from the conventional upper left. Transfer the drawing to the illustration board, using the single-transfer method (2-41). The double-transfer method (2-40) can be used to transfer a line drawing onto a wash surface but not to transfer shaded areas. If all the pencil-shaded areas were transferred onto the illustration board, they would affect the smooth layering of the pigment.

Clean the board again with an eraser, making sure that there is no pencil dust on the surface. A kneaded eraser is good for this, as it picks up dust gently without disturbing the tracing lines and without leaving any crumbs of its own.

Apply the lightest smooth wash on the top of the cube, starting with the clear wash. Apply the wash neatly and crisply and only to the appropriate area. When the first area is completely dry, apply a darker-valued wash to one of the adjacent areas. When that is dry, cover the third side.

Turn the board so that you are always working in horizontal strokes on a tilted, gravity-assisted surface. Brushstrokes should be back and forth, not up and down.

The cube consists of crisp-edged planes, but the cast shadow does not have such a crisp edge. To produce a softer edge, apply the clear wash in a larger area than the shadow is to occupy. After partial drying apply the pigment as a graded wash but do not carry it to the edges of the clear wash (5-9).

Because the more distant surface of the cone consists of two thicknesses of material, it should be darker than the nearer surface. This distant side is covered first. When it is dry, a graded wash is applied on the near surface. Recognize the constant factors of light and shadow: the highlight, the intermediate values, the core dark, and the reflected light. If you make a conscious effort to look for these on all rounded shapes, they will quickly become obvious and easy to recognize.

The sphere is an excellent practice shape, as it is related to many forms in nature. The clear wash is first applied to the entire surface of the sphere and left to dry partially. Starting

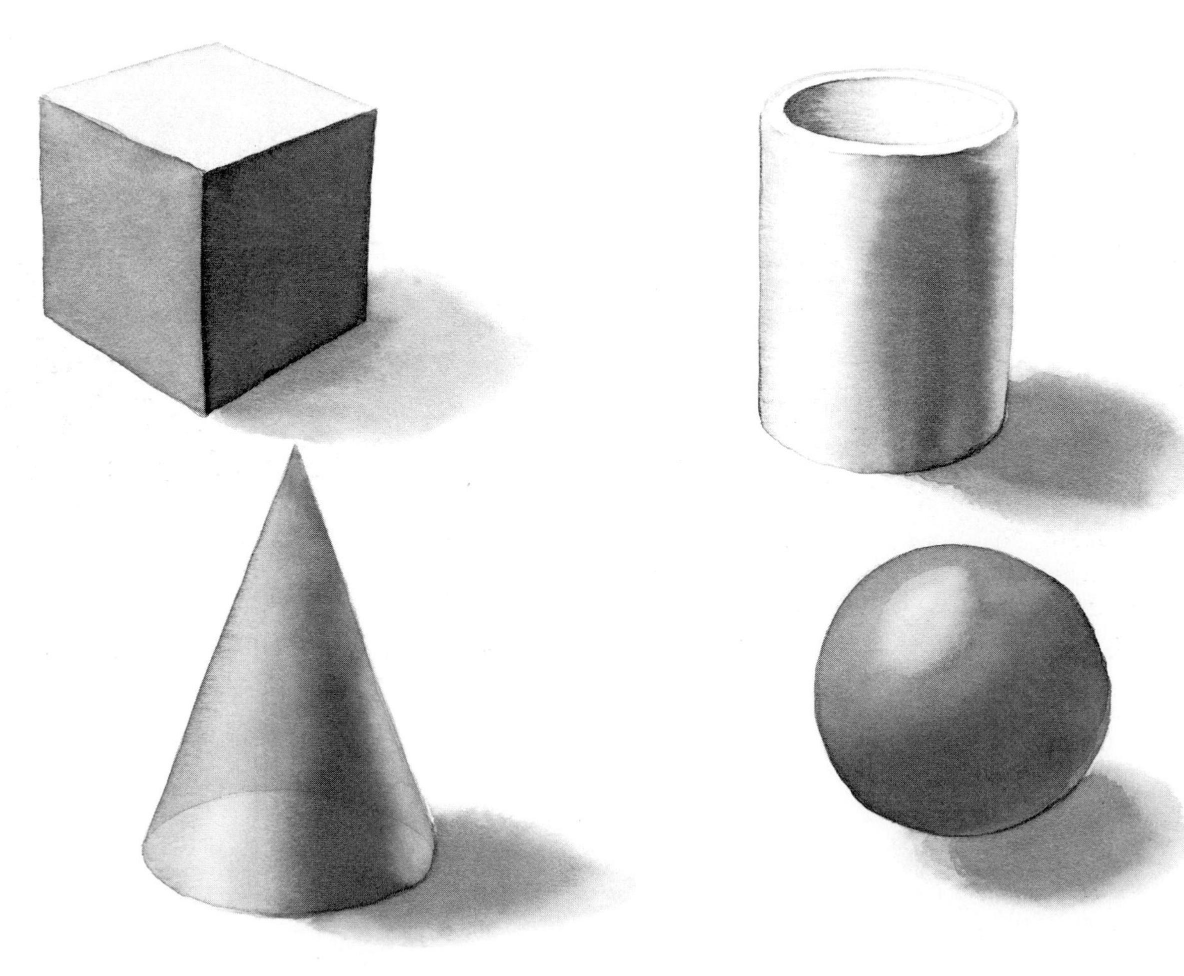

5-9.

with a medium value at the upper circumference, a graded wash is worked in a semicircular manner. More pigment is added in the core-dark area, and a more dilute mixture is used for the reflected-light area. The hair dryer can be used to advantage to "freeze" or stop the action of the wash. Add a soft-edged cast shadow. A small sponge or tissue can be used to blot the edges. The darkest part of the shadow is closest to the sphere.

Finishing

After rendering an illustration in wash look at it with a critical eye. There may be something that can be improved. The deep darks may need to be enhanced with paint in a drybrush technique. A sparkle of white paint in the darkest area may add depth. An edge may need a sparkly highlight of white watercolor or opaque paint. Ink or a carbon or graphite pencil may be used to make details crisp or darker. A knife blade can be used to scratch highlights. A Pink Pearl eraser can be used to soften and lighten certain areas. An electric eraser may be helpful in removing darks or in lightening the value. An eraser, however, cannot remove particles of pigment that have settled in the roughness of cold-press board, so the erasures will not be smooth. Many of these techniques will mar the surface of the board, so they should be done only after all the rendering is complete. White paint must not be disturbed after it has been applied, as it will turn muddy.

5-10. Jane Rady. Leaves. Wash and resist on cold-press illustration board.

Resists

There are a number of commercial resist products, such as Miskit or Maskoid, that block the absorption of the pigment on the board. You can also use thinned rubber cement, which is the least expensive but has no color indicator, making it difficult to see which area is covered.

Dip your brush in liquid soap before dipping it in the resist, and it will be easier to clean. Apply the resist to the area that you wish to leave white. Lay the washes on and, when they are dry, remove the resist by gently rubbing the surface with a finger or rubber-cement "pickup." This leaves a white surface, which can then be left white or painted. A resist can be applied between washes, leaving one part of an area lighter than another. This method of isolating white areas produces crisp edges (5-10).

A resist can be laid over the entire area of the drawing, and the entire board painted with a smooth or graded wash. The resist is then removed and the specimen painted. This method is particularly helpful in rendering a light subject on a darker background. It is also good for special continuous-background effects (5-11).

5-11. Susan Russell. Crab claw. Wash and resist on cold-press illustration board.

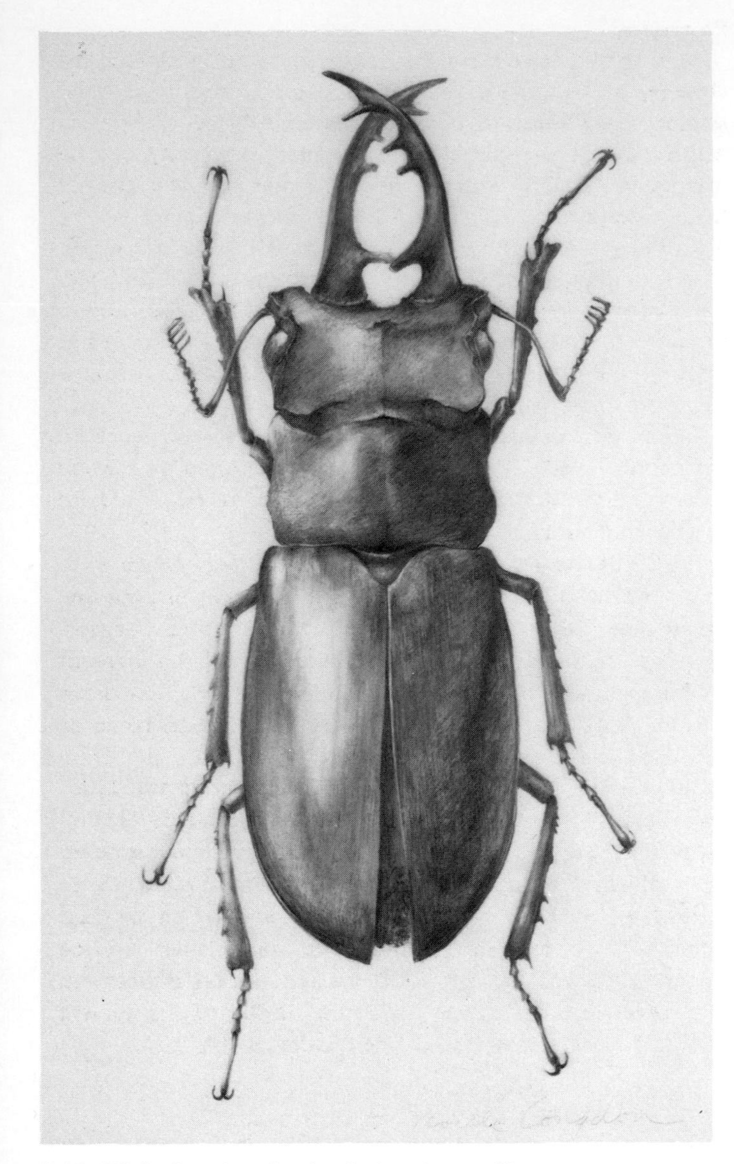

5-12. Nöelle Congdon. Beetle. Carbon dust on film.

CARBON-DUST TECHNIQUE

The carbon-dust technique (5-12) is sometimes called the Brödel technique after its inventor, Max Brödel, the "Father of Medical Illustration" and director of the first school of medical illustration. Many old medical books are enhanced with Mr. Brödel's sensitive yet strong work.

This technique, which is a combination of drawing and painting, yields realistic effects quite quickly and can be used to render fine detail. It is useful for special effects, such as smooth surfaces, brilliant highlights, and strong contrasts. It is generally best to protect these drawings with a mat insert (5-13) or a stable acetate cover. If extensive handling is expected, they can be sprayed with mat fixative, although the spray may alter the integrity of the tones. When using fixative on a drawing, always test first, as brands vary and may spot or produce a white film on your drawing.

Following are some recommendations for carbon-dust materials (5-14).

1. Ground: kid-finish three-ply Strathmore to start (other grounds are discussed later in this section)
2. Pencils: Wolff's carbon drawing pencils 3B (soft), 2B, B, HB, H (hard)
3. Brushes: large, fluffy squirrel or camel "sky" brush or soft, short-bristle flat brush; several old (worn-down) oil-painting brushes or other flat, short-bristled brushes
4. Erasers: kneaded eraser, pink typewriter eraser in pencil shape, plastic or soap eraser
5. Sandpaper pad
6. Tortillons: #1, #3 (paper stumps)
7. Chamois skin
8. Knife: Bard-Parker surgical scalpel with blade #15 or X-acto knife with blade #16

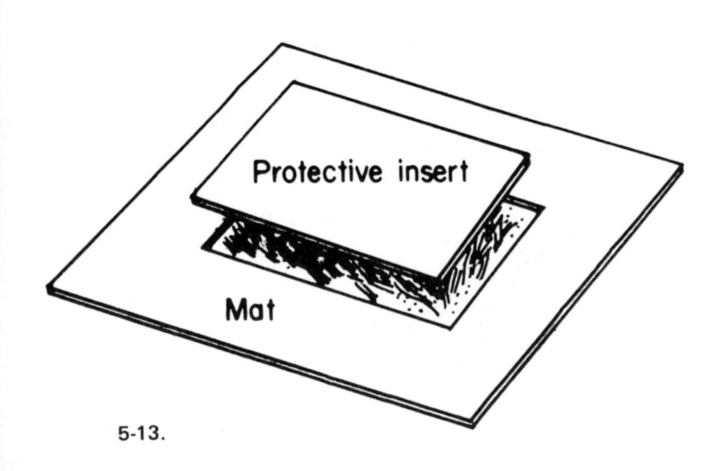

5-13.

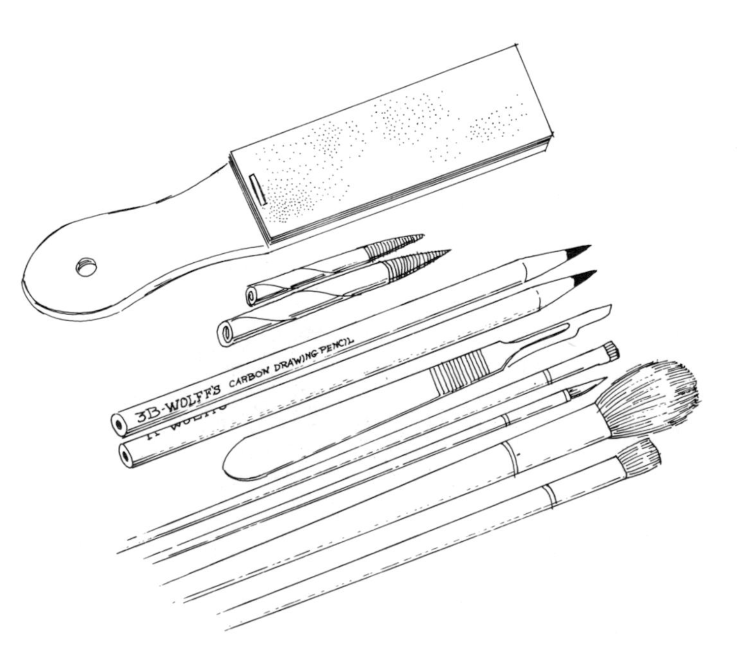

5-14.

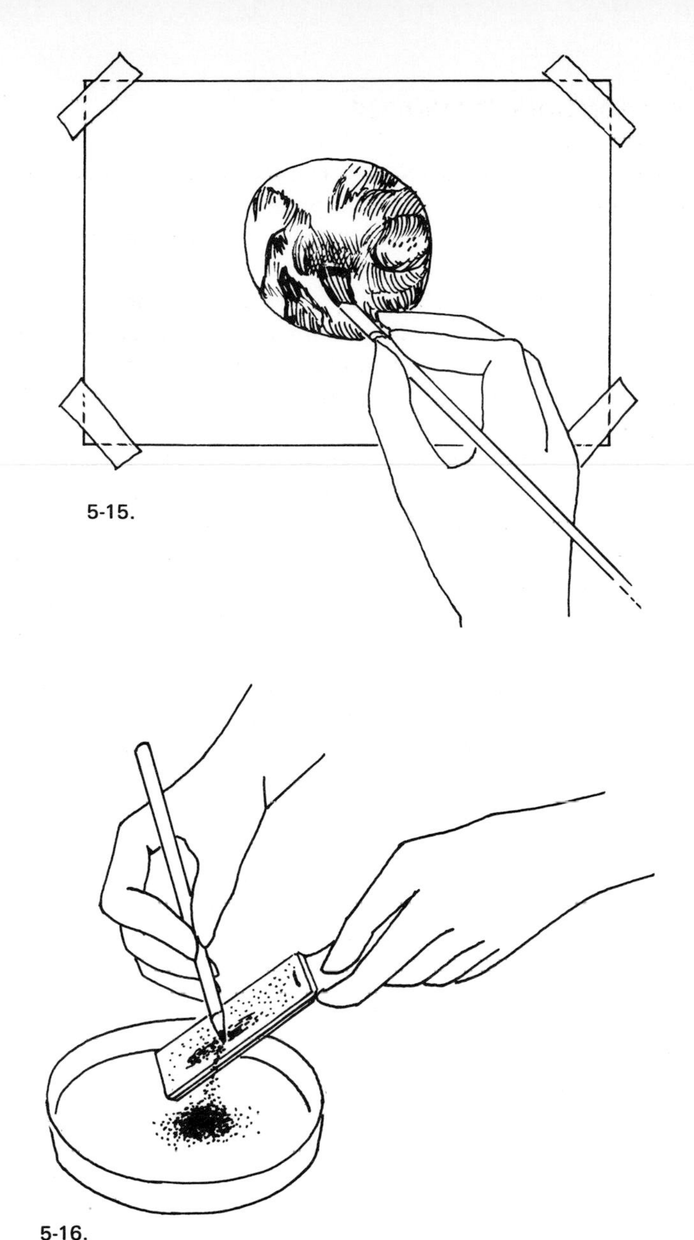

5-15.

5-16.

Procedure

With carbon dust more than with any other technique the virginity of the ground surface must be preserved. Any mar, scratch, hand moisture, or oil will adversely affect the smooth application of the carbon-dust particles. Erase the ground surface thoroughly with an art-gum eraser or eraser crumbs before doing the transfer. Never touch the center of the ground with your fingers or hand: handle it by the edges. When working on your drawing, keep a tissue under your hand. On a large drawing a paper mask can be taped in place, exposing only the area that you are working on. If it is taped in place, it will not rub against the drawing and can be shifted as the drawing develops (5-15).

Carbon dust or sauce is made by rubbing a carbon pencil on a sandpaper pad. The resultant dust is captured on a white paper or dish. A supply can be made ahead of time and kept in a covered container (5-16).

The preliminary drawing should be fairly tightly drawn, with values indicated (5-17). It should be done in line shading rather than in smooth tones, as the latter do not transfer well. The lines also indicate the general direction for the shape of the subject, and, although nearly obliterated in the final drawing, the little that is left lends added depth and character to the shading.

Dip the big, fluffy brush into the sauce and gently brush and shake the excess onto a paper towel or chamois. If you do not remove the excess dust, you may spot your ground. Paint the dust onto the drawing in sweeping strokes. Cover the entire drawing surface, very gradually establishing the main values. Do not try to do it all at once. Many layers produce smoother tones. Do not scrub the dust in: let it float on. This procedure establishes the overall shape of the subject without regard to detail or surface texture (5-18).

5-17. Jessie Phillips Pearson. Carbon dust on ledger paper.

5-18. Jessie Phillips Pearson. Carbon dust on video paper.

5-19. Jessie Phillips Pearson. Carbon dust on video paper.

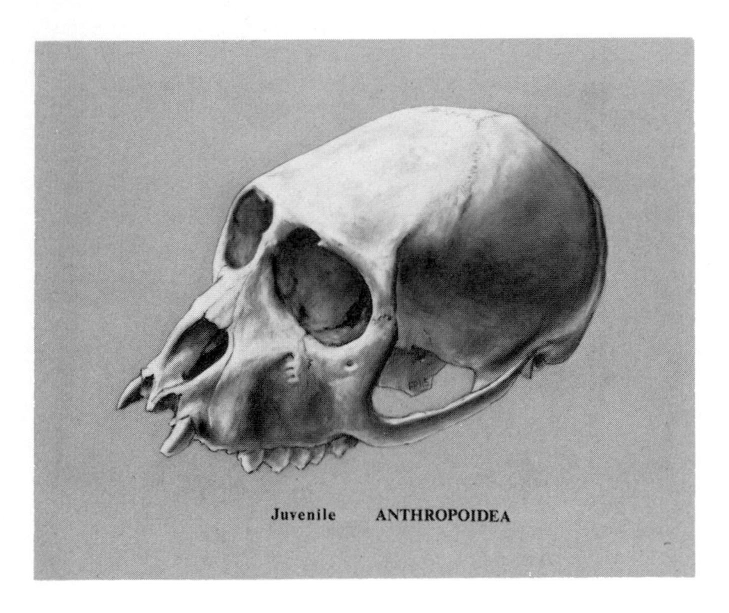

5-20. Anton Friis. Carbon dust and backpainting on film.

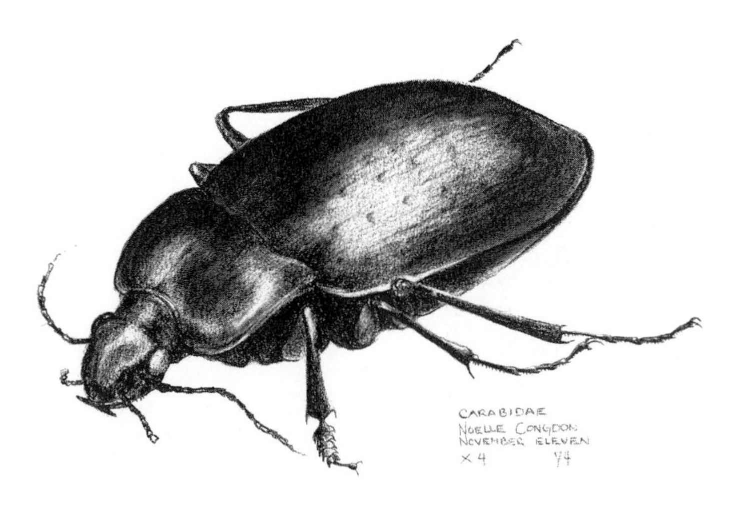

5-21. Nöelle Congdon. Carbon pencil on ledger paper.

When the overall contours have been defined, put in the darkest areas. These may be "painted" in with one of the flat brushes and stomped (rubbed in with the tortillon). The dust may be reapplied several times to get a deep dark. Lighter values are made by removing the carbon dust lightly and delicately with a clean chamois or more definitely with a kneaded eraser. Choosing the appropriate brushes, continue to develop values as you would in any medium, working over the entire drawing and comparing lights, darks, and relative values.

Tortillons or stomps are blotting papers that are rolled to a point. The rolled edges may have to be trimmed to form a smooth surface. When dirty, they may be trimmed with a sharp edge to renew them. Tortillons are used to smooth and extend tonal areas. To create a dark tone, an area can be penciled, stomped with a tortillon, repenciled, and restomped. Tortillons can be loaded with carbon from a penciled area and used as "pencils."

The kneaded eraser can be pointed to remove small highlights. Harder erasers can be cut into sharp points or chisels to remove points or lines of light crisply. If an edge becomes fuzzy, a piece of paper (straight or cut to fit) can be placed over the edge of the drawing as a protector, and the abutting edge erased.

Finishing

When the drawing has progressed as far as is possible with these procedures, you may add some other touches. India ink or graphite pencils may be used to crisp and blacken dark areas. White acrylic paint can be deftly touched on a highlight area. A knife can be used to scrape away the surface carbon. These effects should be added after all the carbon dust has been distributed to your satisfaction, as they prohibit smooth acceptance of additional carbon dust and are irreversible (5-19).

Grounds

The carbon-dust technique requires a ground with a fine tooth or grain to hold the particles of dust. Any good-quality paper with these characteristics will work, although the traditional starch-coated surface has some advantages over noncoated surfaces. The quality and availability of such papers vary. The original #00 Ross board, which is no longer available, can be replaced by Crescent 220, LineKote illustration board, Anjax Video-media paper, Colormatch paper, or dull 100-pound Champion Wedgewood coated cover. These papers must be mounted on a fairly sturdy mounting board, as they are fragile and crack easily. The unique quality of the starch surface allows you to scrape the carbon dust and starch away gently with a knife blade, producing brilliant, crisp highlights. Scraping, of course, mars the surface of the paper and must be done with discretion.

Backpainting

Another type of ground that is particularly versatile is a mat or frosted acetate. The quality of various brands also differs, but I have used both Cronaflex and Bienfag Frosted Protectoid with success. You can skip the entire transfer process simply by laying the acetate mat side up over your original sketch and tracing it.

One of the most effective and dramatic possibilities of frosted acetate is backpainting. After the subject is rendered, the back (smooth side) of the acetate is painted with white acrylic paint under the drawing. The acrylic paint is flexible and does not chip or crack. Several coats are necessary to make it perfectly opaque with no shadowing. The drawing is then mounted on a gray board, making the subject (5-20) stand out in dramatic relief. The principle is that the white paint represents the value of the subject in full light and the carbon dust represents the shadows on the subject. Do not backpaint the cast shadow. This principle can also be used with color (see chapter 6).

Direct pencil rendering

Carbon pencils may be effectively used directly on the ground, with little or no stomping with tortillons (5-21). This is a fast method and produces a more casual result.

GRAPHITE-PENCIL DRAWING

Pencil shading is easily adapted to scientific illustration and ideally suited to rendering fine detail. The materials are commonplace, easily available, and simple to work with (5-22). It is a flexible technique, lending itself readily to alterations and corrections.

The principal difference lies in the stroking. Stroke marks, a characteristic of pencil drawing, must be limited to backgrounds, cast shadows, and nonessential parts of the drawing. In a scientific illustration they may be misinterpreted as the surface texture of the specimen and thus should be avoided in these areas. Practice and the use of a different method are necessary in order to obtain a smooth tone over a large area.

5-22. Trudy Nicholson. California brown pelican. 9H graphite pencil on media board.

Materials

Following are some recommendations for pencil-drawing materials.

1. Ground: three-ply smooth-finish Strathmore; Cronaflex, Bruning Sure-Scale frosted acetate
2. Pencils: Eagle Turquoise or Venus graphite 4H (hard), 2H, HB, 2B, 4B (soft); Mars-Duralar K-1 (soft), K-4 (hard)
3. Tortillons: #1, #2, #3
4. Erasers: kneaded eraser, Eberhard-Faber Kleen-off #41, Weldon Roberts wizard vinyl-pencil eraser
5. Scrapers: Bard-Parker scalpel with blade #15, X-acto knife with #16 blade

It is necessary to use a ground with some tooth to capture and hold the graphite in order to build up values. The smooth-surface Strathmore paper has a slight tooth—it is not quite so smooth as plate finish—but does not have a grain. Make sure to purchase frosted acetate that does not have a grain and is free from scratches, pits, or other imperfections.

Procedure

To transfer the preliminary drawing to the paper, use either the single- or the double-transfer method (2-40, 2-41). Check the transfer for lost or strayed lines. It is usually sufficient to transfer an unrendered outline drawing. Remove any extra graphite gently with a kneaded eraser. Transferring is not necessary with the acetate film, as the sketch is laid under it and traced directly. The preliminary sketch should, however, be sprayed with fixative to prevent it from being transferred to the back of the film. Trace the drawing lightly on the acetate with a 2H or K-5 pencil, remove the preliminary drawing, and substitute several sheets of ungrained typing paper. Pencil strokes on acetate pick up any underlying paper grain, which is usually undesirable.

The most difficult part of the technique is to lay in the pencil strokes smoothly without revealing any pencil lines. A very sharp pencil is the prime requisite. Use a hand-held pencil sharpener or sandpaper pad frequently and sharpen the pencil further by rubbing it on a piece of paper—linen-finish notepaper is good.

The pencil is held at a low angle (nearly parallel) to the paper and stroked back and forth with equidistant strokes at a constant pressure. Use the middle finger as a "slider" to control the pressure of the pencil. The harder the pressure of the pencil, the darker the tone. The area may have to be covered several times to obtain the desired value. The pencil must be sharpened constantly. Tape a mask or a piece of paper under your hand for protection against moisture.

The entire drawing can be worked in this direct manner, using a single pencil (2B or K-1), without using tortillons, removing any graphite with the eraser, or cutting with the knife blade. The success or failure of a pencil rendering depends on your control of the pencil strokes. The direct strokes must be consistently close together, equal in pressure, and long and smooth, not short or curly.

As with most other techniques, the main shapes or contours of the subject are built up first. After the primary shapes are established, details and surface markings are developed by adding graphite to and removing it from the drawing. It is important during this process not to disturb the graphite already laid down. A paper mask is taped over the drawing, revealing only the area that is being worked. Taping it keeps your hand off the drawing and also keeps the mask from sliding over the drawing, shifting or softening tones.

Most of the values are put in with the softer 2B or K-1 pencils. Outlines and detail are added later with the harder 2H or K-4 pencils. The softer pencils do not "hold" on top of the harder pencils. Duralar pencils produce the blackest tones.

The kneaded eraser can be pointed to clean the area surrounding the drawing. It can be used to blend and lighten areas. The plastic eraser (Kleen-off or vinyl) is smooth and nonabrasive—rough ink eraser would abrade the ground. It should be trimmed to a point on one end and to a chisel edge on the other (5-23). It is used to lift out tone in small, specific points or lines. The knife can be used to pick out whites. If used with a light touch, you can repencil the area. The tortillons can smooth and extend tones when used with long, smooth, delicate strokes (never short, jerky, or circular).

After the penciling is complete, white acrylic paint can be added with a pen or brush for further emphasis. When using the film, the entire back of the drawing (or parts of it) can be opaqued with several coats of white acrylic paint and mounted on a gray board for further emphasis.

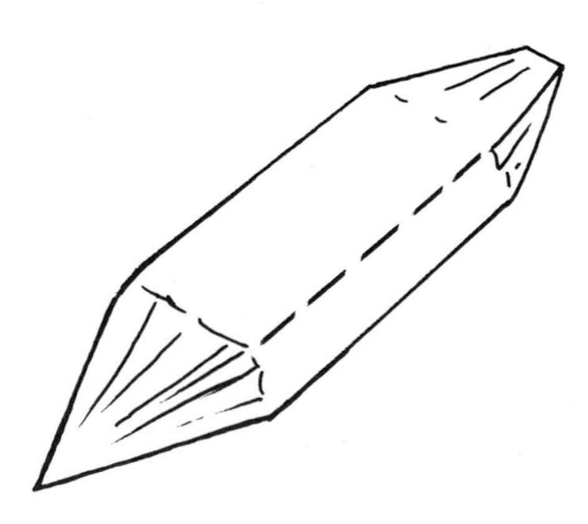

5-23.

AIRBRUSH TECHNIQUE

The airbrush is a sophisticated adjustable tool that conveys dilute paint to a ground in a very fine spray propelled by controlled air pressure (5-24). Photo retouchers use the airbrush to alter photographs, diminishing defects and unnecessary detail, and to fade or opaque backgrounds. The artist uses the airbrush as a brush or pencil to render values and to draw lines or spots. While the pencil is at its best in rendering detail, the airbrush really shines in producing smooth and graded areas of tone. These areas are strokeless and brushless in quality and may be rendered very quickly. Subtle blending and crisp edges are both quick and easy to obtain after some practice.

The main deterrents to the use of the airbrush are the artist's innate mistrust of mechanical instruments and the initial expense and nonportability of the equipment. After the necessary investment is made and a permanent part of the studio reserved for it, you should find the airbrush a delight for rendering either by itself or in combination with other techniques. You can produce smooth and graded tones and crisp edges in record time, a feat that would be impossible with any other medium.

5-24. Joel Ito. *Lemur catta* (ring-tailed lemur). Airbrush and watercolor on film.

Materials

Following are some recommendations for airbrush materials.

1. Airbrush: Paasche AB or Thayer & Chandler A double-action airbrush
2. Pigment: lampblack watercolor (tube)
3. Ground: hot-press illustration board (smooth); three-ply bristol board; Cronaflex, Bruning Sure Scale, Bienfang Protectoid frosted acetate
4. Eraser: Pink Pearl, ink eraser, Rub-Kleen soft-pencil eraser
5. Scraper: Bard-Parker scalpel with #15 blade, X-acto knife with #16 blade
6. Frisket: commercial frisket paper, liquid frisket, .005 acetate
7. Brush: Winsor & Newton series 7, #00, #0, #1

The double-action airbrush is equipped with two valves and designed to do fine and delicate work as well as to cover broad areas. There are many airbrushes on the market, which vary greatly in price and capabilities. A thorough study of the advantages of each in relation to your needs is advisable before making this investment.

The airbrush is controlled by a lever that is manipulated by the index finger. The up-and-down movement of the lever controls the pressure of the spray. The backward-and-forward movement of the lever controls the size of the spray area. A short downward and a short backward movement of the lever produce a fine line with the airbrush close to the surface of the ground (5-25). A long downward movement and a long backward movement of the lever conversely produce a wide spray with the airbrush further from the ground (5-26). By coordinating these two actions one can paint anything from a small dot or line to a large flat area.

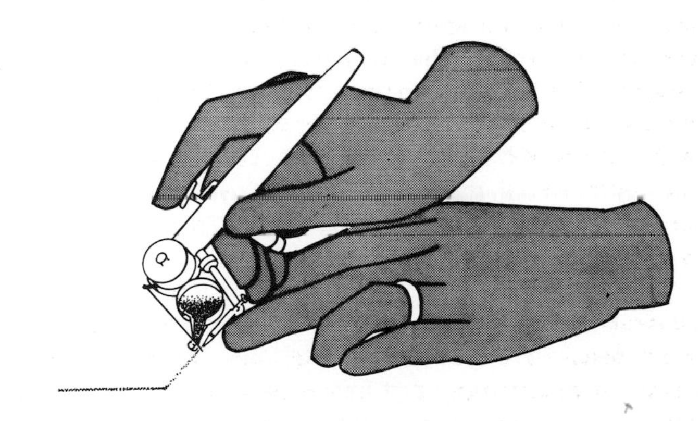

5-25.

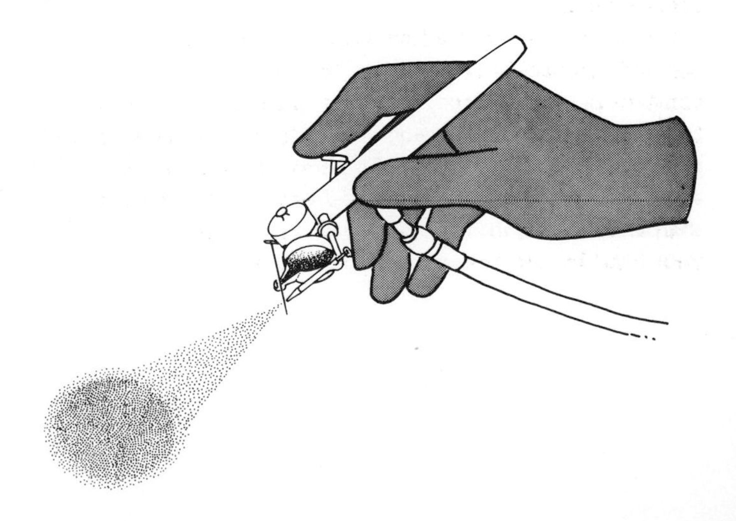

5-26.

The airbrush must be kept clean, taking care that no dried paint accumulates in the air cap or in the needle opening to mar the delicacy of the spray. Run clear water through it often as you work and flush it thoroughly after using. Holding it briefly in an ultrasonic cleaner after using it ensures the removal of all particles of dried paint. The needle should be checked often. If it is bent, the quality of the spray will be affected.

Air compressors are generally somewhat noisy but are more portable than carbon-dioxide cylinders, which must be replenished when empty. Base your selection on your working situation and on the amount of money that you can spend, as types vary considerably in price.

Frosted acetates allow you to omit the transfer stage, to scrape for highlights, and to backpaint and mount on gray board for further highlighting and emphasis. Smooth illustration board is also a satisfactory ground. After it has been covered with the droplets from the airbrush, it will have a toothy quality that will accept the carbon-dust and drybrush techniques.

Procedures

The hard, mechanical appearance that is usually associated with the airbrush can be avoided by using the casual-control mixed-media approach that has been perfected by Joel Ito. Both this technique and the more conventional tight-frisketing technique are described below.

Casual-control style

Trace the drawing very lightly on the frosted acetate, using a gray Prismacolor carbon or lead pencil. Mix the paint on a palette, as with any watercolor technique, and transfer the diluted paint to the airbrush cup with a small watercolor brush. Put only a small amount of paint in the airbrush each time to prevent overspraying. A full cup and the tendency to spray until the cup is empty encourage you to darken too quickly.

Test the spray on a scrap of paper before every stroke for tone and quality. Any malfunction or spitting can be corrected before touching the spray to the paper. Thick dilutions of paint are mixed for the darker values, and thin dilutions for the lighter values. A solution is generally too thick if the spray comes out grainy or spotty and too thin if it runs when it hits the ground.

The airbrush must be kept in motion at all times. It must never be held in one spot, or it will blow holes in the layers of paint. While the airbrush must be very close to the ground to render fine detail, it must never touch, as this would mar the surface and blow the paint away.

In airbrushing minute droplets of dilute pigment strike the ground in a nearly dry state, as they are almost instantly dried in the airstream. Under a magnifying glass the droplets or spots of paint are apparent (5-27). For this reason it is wise to start out with a lighter dilution than desired. The area should be covered many times, gradually building up the tones and establishing the main contours. Applying more and more

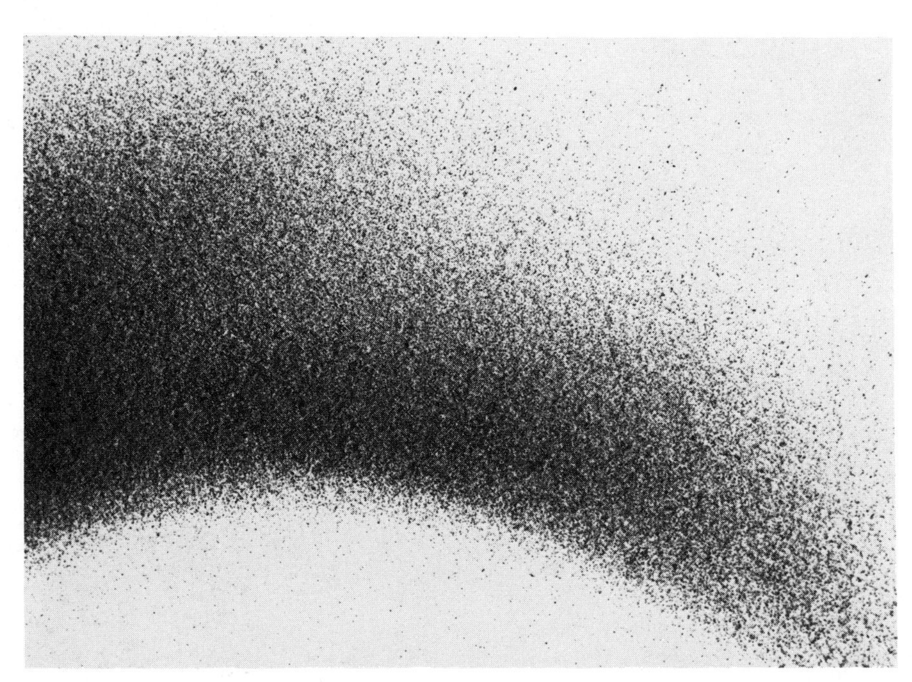

5-27.

droplets of paint gives you the depth of tone that you want, ranging from a faint veil to an opaque area.

If an area is too dark or unsatisfactory for any reason, it can be lightened by rubbing gently with a Pink Pearl or ink eraser. Small areas can be wiped off completely with a damp tissue or moist Q-tip if you are using acetate film.

Detail

Most airbrush renderings seem soft without the addition of accents for texture and modeling. These may be added with a #00 series 7 Winsor & Newton watercolor brush. Use a rather dry brush and a light touch to avoid picking up the layers of paint. Black accents may be added with a Gillotte #290 pen and ink. Small, crisp highlights and sparkling reflections can be obtained with white opaque watercolor or by scratching with a sharp knife. Reapply the airbrush to establish the main darks firmly. Add final lines for emphasis; erase small areas for subtle highlights. To spotlight certain areas, you can airbrush or paint with opaque white paint on the reverse side of the acetate. Mount on a gray or white board, selecting the value that sets off the drawing to best advantage (5-28).

You can also do the brushwork first: establish areas of texture and hard edges and then use the airbrush to tie everything together, spraying over the brushwork to establish general light areas and dark shadows. Brushed-on accents can be added afterward.

a

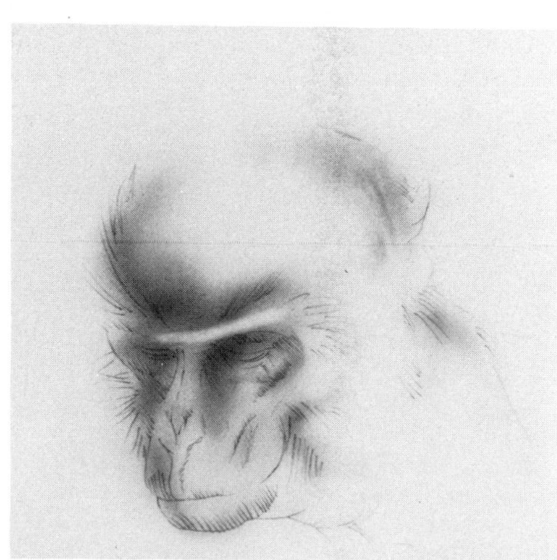

b

c

d

e

5-28. Joel Ito. *Macaca mulatta* (rhesus monkey). Airbrush on film.

then transferred to another tape to be viewed, and the skill of the particular specialists. With projectables it is safest to test colors and combinations with the equipment that will produce the final products. In print media experience and the advice of a good printer are your best resources.

Four-color-process printing

While only one plate is necessary for a black-and-white line cut or a continuous-tone drawing (see chapter 5), four screened plates are necessary to print a full-color drawing (C-4). The original artwork is photographed with selective filters to screen for each of the three basic colors. From these negatives three plates are made: the yellow printer (C-5), the cyan (blue-green) printer (C-6), and the magenta (red) printer (C-7). A black printer is added to overcome the muddiness of the dark tones (C-8). Each of these colors is interpreted separately in a dot pattern on a screen or grid in the same manner as for the black dots of the continuous-tone screen. Each colored dot takes its own place in a rosette consisting of one dot of each color (C-9). To avoid superimposing the dots on one another, which results in a moire pattern, each screen is carefully angled from the others at a predetermined degree. An error in the angle of the screen as little as 0.1° can produce a moiré pattern.

In the predominantly yellow area the yellow dots are large. They are small in the area with the least yellow. This same principle of saturation of each color determines the size of each of the color dots in the rosettes. Your eye blends these dots together so that they appear as smooth gradations of color. Look at a color photograph in a printed publication with a magnifying lens. The rosettes formed by the four-color dots are visible. The finer the grid pattern or screen that is used, the more dots or rosettes per square inch and the truer the print is to the original. This is obviously an expensive process, reserved for long-run publications and special considerations.

When doing color work that is to be printed with four-color-process printing, it is best to work with colors and combinations of colors that are similar to those of the three-color printers: a pure yellow for the yellow printer, a bluish red (like carmine) for the magenta printer, and a greenish blue (Prussian blue) for the cyan printer. Black tends to degrade the reproduction, so use it very sparingly. Special process colors are available from any graphic-art dealer. It is good to check with the printer if possible to determine which process colors they are using.

Preseparated color printing

Preseparated or flat color artwork is prepared in black or photographically opaque material and white. The key or main color (generally black) is on illustration board, and the other colors are on registered overlays. This setup is called a mechanical and is camera-ready for the printer.

The main or key-line drawing can be in either line or continuous tone and is either drawn or mounted on illustration board. Three registration marks are placed on different sides of the drawing just outside the area to be printed. The artwork for each secondary color is prepared on a separate overlay of a stable, translucent material such as Mylar or frosted acetate. These overlays are hinged to different edges of the main drawing. Registration marks on the overlays are superimposed on the key-line registration marks (6-1). Registration marks of concentric circles centered by a cross can be purchased in adhesive-backed rolls.

The artwork on the overlays must be light-safe or opaque for the platemaker. This can be done with opaque black ink or red opaquing liquid made especially for the purpose (Parapaque is one brand). If the area is large or if long, smooth edges are important, the color areas on the overlays can be prepared by cutting sheets of adhesive-backed masking film that is light-opaque (Rubylith and Lithoblock are two brand names). The color of the opaquing material does not relate to the color that the area is to be printed. The artist writes the ink color and the desired percentage of that color on the appropriate overlay. A designation of 10% blue would be printed as a light blue; 50% blue would be a medium blue; and 90% blue, a deep blue.

If there are several colors that are separated from each other, they can be put on the same overlay, with the percentages of color clearly directed to the various areas of the overlay (6-2). Printing directions that fall within the printing area can be made with a light blue nonreproducing pencil. If several colors must meet on the finished artwork at a hairline register, they should all be on the same overlay. A tissue overlay is placed over it to indicate the separation of colors (6-3).

Several colors can be combined in the same area: for example, 100% yellow/20% black will produce a grayed yellow, while 60% cyan/100% magenta will produce a purple (6-4). Printing inks are manufactured in six to eight basic colors. These can be mixed according to prearranged formulas to produce about 400 colors. There are charts that show these colors printed on both white and colored stock in different paper finishes. The Pantone color-matching system is commonly used. Good rapport with your printer is important, as each printer has different capabilities and specialities.

Duotone

A duotone is the result of two screen negatives of a single piece of continuous-tone illustration or photography, each printed in a different color (6-5). The screens are at slightly different angles so that the different-colored dots are not superimposed on each other and are photographed with different screen values. One is noncontrasting and contains full detail, and the other is more contrasting. The combination of the two gives the picture more depth of color and definition. The entire illustration can be treated in this way, or the second color can be added to only a specific portion of the plate. This area is indicated by outlining it on a registered overlay.

A third color or tripletone is also possible with a single piece of artwork. For duotones and tripletones the original illustration should be strong in contrasts with well-modeled middle tones and sharp blacks and whites. This method of adding a color can be as instructive as process or full color at a fraction of the cost.

Labeling

Labels, leaders, and body copy are put on the key-line drawing if they are to be in the same color (commonly black). They are put on a registered overlay if the drawing is in full color and it is to be printed. This overlay is not screened but rather photographed separately by the platemaker and printed as a line cut. Labeling for nonprint media (slides, TV, etc.) is put directly on the colored drawing.

COLOR QUALITIES

Every color has five qualities: hue, the name of the color (red, yellow, blue); value, its place on the light-dark scale; chroma, the purity of the color, or absence of white or pastel tendency; brilliance, the vividness or absence of black in the color; and temperature, its warmth or coolness.

Altering the pigment

The qualities of a color can be altered by mixing the pigment with more or less water or with one or more other colors.

Greater dilution of the pigment with water permits it to be spread thinner, allowing more of the white ground to show through and producing a color high in value (C-10).

The addition of white pigment to color heightens the value, increases the opacity or pastel effect, and gives the color more "body" (C-11).

The hue of the color can be shifted by adding one or more colors to it. This also affects the brilliance, chroma, and value of the color (C-12).

A color can be neutralized or grayed by the addition of its complementary color, the one opposite it on the color wheel. The complementary pairs are yellow and violet, orange and blue, and red and green. Equal amounts of each of the complementary pair ideally produce gray (C-13, C-14, C-15).

A color can also be lowered in value or grayed by the addition of black, brown, or gray, although this tends to deaden the color. Brown is used to darken yellow, as black gives it a greenish hue. These colors also neutralize the hue and brilliance, essentially graying the color (C-16).

It is easier to neutralize the brilliance of a color than to increase it, so it is better to start with clear colors and gradually to degrade or gray them. As black is composed of all colors, the addition of each color is a step in the direction of black.

In general warmth is on the orange side of the color wheel, including red and yellow; coolness is on the blue side of the color wheel, surrounded by violet and green. It is conventional to think of cool colors as receding and warm colors as advancing, but there can be many peripheral influences: the total area, brilliance, and surrounding colors affect this "rule." Hue mixing can also produce warm greens and cool yellows.

Local color

Local color is exterior color on a subject without outside influences, just as if it were painted on a flat surface. When starting to paint, select the color that is closest to the local color and modify it with a measured dilution of water and careful addition of other pigments (C-17).

Local color is influenced in many ways. It is most true on a flat, mat surface with good diffuse light. In shadows it is degraded or neutralized and lower in value. In direct light it is neutralized and higher in value. Atmospheric perspective causes local color to lose its brilliance: it becomes more neutral and lighter in value as it recedes from the viewer (C-18).

Reflections from surrounding colors influence local color. Every surface is in effect a mirror, reflecting the neighboring colors. Shiny surfaces show more of the neighboring colors than do dull surfaces. Cast shadows echo the local color of the specimen (C-19).

Textures influence the local color: a dull or fuzzy surface degrades the brilliance and lowers the value of the color, while a shiny surface shows the brilliance and purity of the local color in the light areas and nearly obliterates all color in the highlight areas (C-20).

WATERCOLOR

With watercolors it is possible to obtain smooth, even gradations within a color and from color to color (C-21). Clarity and brilliance as well as subtlety of tone and intricacy of detail are not difficult. Freshness and vitality are characteristic of this water-based medium.

The instructions and materials supplied for wash drawing (see chapter 5) are applicable, with the addition of a range of pigments, which are available in tubes. A minimum basic selection is listed in the left-hand column, with additional colors on the right.

1. Yellow	
cadmium yellow light	yellow ochre
	cadmium yellow medium
	lemon yellow
2. Red	
cadmium red light	cadmium red medium
alizaron crimson	
3. Blue	
ultramarine blue	cerulean blue
	phthalo blue
	Winsor blue
4. Green	
phthalo green	Hooker's green 1
	Hooker's green 2
5. Black	
ivory black	Payne's gray
6. White	
chinese white	
7. burnt sienna	
8. burnt umber	

Using your largest pointed brush and cold-press illustration board, make some smooth and graded washes similar to those in chapter 5. Remember to dampen the surface with clear water first and to allow it to soak in until the grain of the board is visible above the water. When you feel comfortable with your materials, start with one color and gradually mix in another until you can manage a smooth transition from one color to another. Start with a pure color and gradually gray it by adding its complementary. Gray the same color, using brown or black, and notice the value reduction. Make a glaze by gently flowing a second color over a dried area.

Grade a wash by gradually adding white. Notice how it differs from a graded wash in which the ground is used as the white. The translucency that is so characteristic of a fresh watercolor drawing is based on the white of the ground. Adding white to the color builds up the body of the color and gives a different effect.

Rendering the drawing

Transfer the drawing using the single-transfer method described in chapter 2 (2-41) so that there is the minimum amount of pencil on the final ground.

Arrange a small amount of pigment from each tube around the periphery of your white china or enamel palette. Have nearby a large water container filled with clean water, your brushes, and tissues. In front of you are your preliminary drawing and your specimen. Refer to them frequently to refresh your memory and to correct, alter, and refine the drawing as the painting develops. If it is not possible to have the specimen with you, your color notes should be complete and precise.

Select from your palette the color closest to the main local color of the specimen. Alter it to get the precise local color. Use as few colors as possible, since the use of additional colors tends to muddy the hue. Dampen only the area that you wish to cover on the first pass. Cover that area with color, grading it for value and brilliance. This develops the main contours of the subject without regard for details. Make subsequent passes of wash after the area is completely dry until the principal shape is established. The structural details, texture, and surface-color pattern may now be worked up, gradually adding more intricate detail and darker darks.

For reflective sparkles, highlights, and detail drybrush opaque white can be added. Add it carefully as the very last step so as not to mix with or pick up the underlying colors. Crisp ink lines can be added for emphasis. A Pink Pearl eraser, used carefully, can lighten values or soften colors.

Friskets

Liquid frisketing material (see chapter 5) can be used as a resist to leave small areas of white ground that can be left white or painted with another unsullied color (C-22). When painting a light-colored specimen on a white ground, it is difficult to obtain enough contrast between the specimen and the ground and still to render accurate color. Liquid frisket can be painted over the entire area to be covered by the specimen, and the background, including the cast shadow if there is one, can be painted in a contrasting value in a smooth or graded wash. When the frisket is rubbed off, it leaves a white surface on which to render the subject so that it contrasts with the background. The background should not compete with the subject in either intensity of color, importance of detail, or design. This method can also be used to render smooth and uninterrupted backgrounds that are appropriate for the subject, such as sky, water, or surrounding tissues (C-23).

ACRYLICS

Acrylics give results similar to watercolors and can be handled in much the same way (C-24). They do not have the translucent look that is characteristic of watercolors, but they do offer two additional advantages: succeeding washes do not affect the underlying pigments, allowing more buildup of pigments and values; and it is possible to work on a colored ground and to build values lighter and darker than that of the ground.

The area on the cold-press illustration board that is to be painted is wetted down first, as with watercolors. Unlike watercolors, the colors are mixed in separate thin mixtures. A small brushful of acrylic mat medium is added to keep the paint smooth and to allow it to bond well with the board. Successive thin washes are applied, as with watercolors, making sure that the surface is completely dry before applying new washes. The undercolors will never "pick up" from the board as increasing numbers of washes are applied. The colors build up rapidly with each wash, and some marvelous transparent effects can be achieved with this method. If a wash gets too dark, a very thin white wash can be used to tone it down. It can also be used to unify several colored areas.

When using colored illustration board as the ground, the board color can function as part of the tone of the drawing, showing through in selected areas. The shaded areas will be darker than the board, the lighter values lighter. Transparent washes are applied first, and, as your drawing progresses, you can use more opaque washes to build up the solid areas. Again, a thin white wash or a tinted color wash can be added overall to pull the entire subject together. Opaque white highlights are added last. Permanent-white designer's gouache is good for very bright highlights. Acrylic white is more transparent and gives more subtle highlights.

Using a black or very dark ground, you can produce a luminous and translucent quality. The center of the subject, which you are looking directly through, is rendered with little or no light pigment; the edges of the subject, which are viewed obliquely or end on, are rendered with the lightest pigment (C-25).

COLOR PENCILS

Color pencils are familiar, comfortable tools to work with,

but they do not give you the smooth, even tones that you can obtain with a wet medium, although richness of pigment and crispness of detail are possible (C-26). Color pencils are particularly useful for sketching on location and for making color notes when it is not possible to take the specimen to the studio.

It is necessary to start with a basic set of at least 12 colors. As it is not easy to make new colors by blending, you may want to purchase additional pencils for each new specimen so that you can match its colors more nearly. Prismacolor, Conte, Castell, and Eagle Verithin are good brands of color pencils. Prismacolor has the most vivid pigment but loses its point more quickly. Eagle Verithin is a harder pencil and keeps its sharp point for a longer time. A Pink Pearl or plastic eraser and an X-acto knife or scalpel are needed. The ground should have a randomly patterned tooth or roughness. Kid-finish Strathmore and cold-press illustration board are good basic grounds. Coated papers give you another option, allowing the use of a blade to crisp up the edges and to remove the laid-on pencil, revealing the white or colored ground, which can then be treated with another color.

The colors are laid in with very short, separate strokes, using very sharp pencils. This is similar to the blending in pointillism and to the screening of a full-color illustration for printing, in which the individually colored dots are blended by the eye. The stroke marks should be so crisp, small, and close together that they are not individually obvious. They can be applied as parallel lines, at right angles, or as a random mosaic. Do not use a tortillon for blending, as it will sully the brilliance of the colors.

The pigment adheres first to the peaks of the grain texture. Kid-finish paper and cold-press board, with its coarser texture, are more satisfactory for larger drawings, and coated papers, with finer textures, are more suited to small drawings. Video paper and Line-Kote board are two coated grounds that have a fine texture. Color-aid and Color-match papers have a very fine surface so that less pencil is caught in the grain.

When the texture of the ground becomes saturated with pencil material, no more color can be added, which means that the vividness cannot be intensified or the hue altered. The correct colors must therefore be applied from the very first strokes. Gently rubbing with an eraser or scraping with a blade may roughen the surface so that it will take more color. It is possible to add highlights of opaque paint to the finished drawing.

As with acrylics, color pencils can be used on black, dark, or other colored grounds. Using the black or dark ground, you can achieve a phosphorescent translucence that is not possible on a white or light ground. The very light parts must be undercoated first with white pencil before the color is laid on, or the value will not be light enough. Allow the ground to show through in all but the lightest areas, leaving the most translucent areas without any pigment to enhance the luminous effect. Using permanent-white designers' gouache for the most intense highlights further increases the iridescent effect (C-27).

AIRBRUSHING

To apply color with the airbrush, use the techniques described in chapter 5 (C-28). Use Winsor & Newton transparent watercolors or any other good brand. The range of colors listed in this chapter under watercolor makes a good basic palette to start with. Any smooth ground is satisfactory: hot-press illustration board, three-ply bristol board, scratchboard, or the dull side of mat acetate. With the last two it is possible to remove paint from small areas with a blade. You can use cold-press board or three-ply kid-finish paper if you wish the texture or tooth to show through.

The drawing should be transferred or traced very lightly onto the ground with graphite or color pencil—only enough to show the main outlines. A dark, detailed transfer will not be covered by the droplets from the airbrush and cannot be erased. If there is to be a crisp outline around the drawing, liquid frisket can be painted around the drawing, with an overlapping paper mask covering the rest of the ground. Paper frisketing may also be used if you take care not to cut into the ground. Any crisp, small details within the drawing can be frisketed at this time.

The colors are sprayed on in the same order in which the plates are printed: yellow, red, and blue. Put only a brushful of dilute pigment in the cup at a time. This helps prevent overspraying, and you need not clean the cup between colors. Apply the color lightly the first time that you cover the area and gradually darken it. The succession of colors may be repeated until the amount of color is correct. Black or grayed color is added last. At this point the airbrushing is complete but has a rather hazy appearance. To crisp up and further define the edges and the details, crosshatch with a conventional brush, using watercolor or opaque paint. The brush should be quite dry to avoid picking up the underlying droplets of paint. Color pencils can also be used at this stage for accenting and outlining. On scratchboard and mat acetate a blade can be used gently and crisply to remove the surface paint and to leave white accents. With mat acetate you can also backpaint highlights with light or white opaque paint. The drawing can then be mounted on a board of any color. A cast shadow can be airbrushed on the background board (C-29).

MIXED MEDIA

There is practically no limit to the combinations of media that can be used together (C-30). If it works, do it. There is, however, a general order in which certain media can be applied. A drybrush technique employing gouache, acrylic, or watercolor can be used over color pencil, transparent watercolor, or airbrush if applied carefully to avoid picking up the underlayers. Color pencil can be used over transparent watercolor, airbrush, and acrylic, and it often effectively ties together or unifies a drawing or parts of it. Watercolor does not adhere to the waxy color-pencil surface or to the sealed acrylic surface and picks up an airbrush surface. Graphite and ink can be used over any of these media, but be careful to avoid harshness.

OTHER OPTIONS

The traditional color techniques for scientific illustration are covered here, but there are other color methods that can be used to produce scientific illustrations that are uniquely yours. Transparent liquid watercolors (Dr. Ph. Martin's dyes, Luma watercolors) are very intense and come in many mixable colors. Subsequent selective application of laundry bleach, which eradicates them, can produce dramatic and realistic effects. Felt-tip pens come in many colors and widths and can be used in a less than casual manner. Opaque papers such as Color-aid and Pantone produce smooth mat areas of brilliant or subtle color. They can be cut out and assembled as a collage or form a background for other techniques. Sheets of translucent color film are available under brand names such as Letracolor, Colotone, Cello-Tak, and Zipatone. They are available in many colors and in various percentages of each color. Silk-screening is a flexible technique (see chapter 10), and there are the old faithfuls, oils, pastels, and crayons. Your own original talent and creativity in concert with a new combination of techniques and materials may make you famous in the scientific community. There are no limits to the imagination!

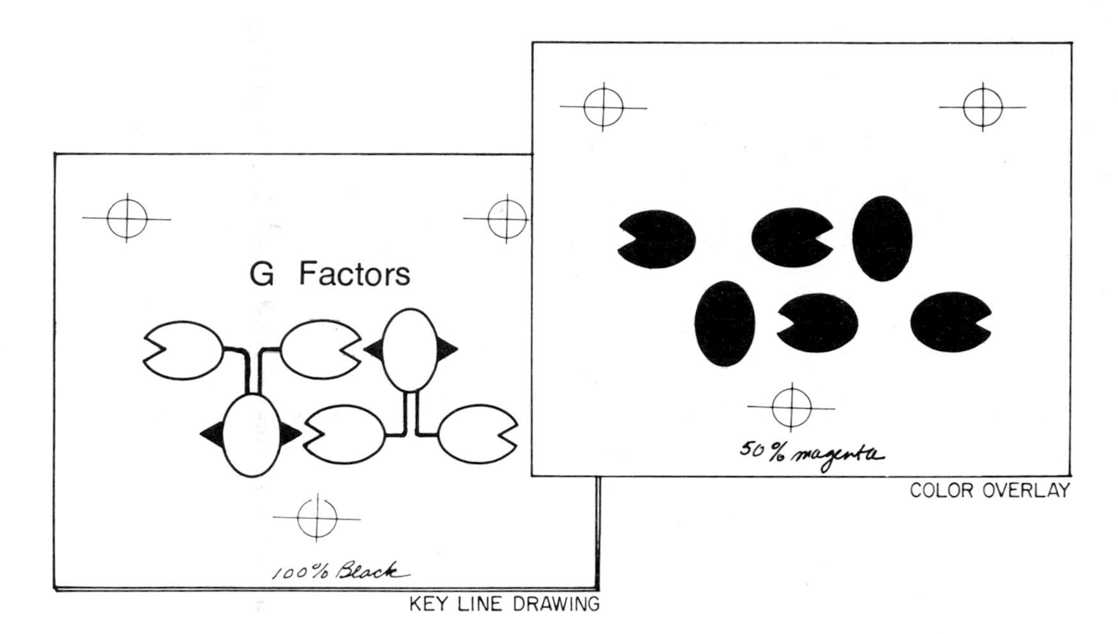
G Factors
100% Black
KEY LINE DRAWING
50% magenta
COLOR OVERLAY

6-1.

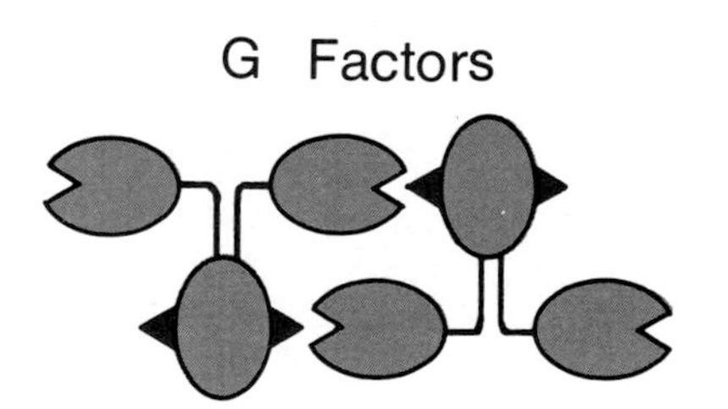
G Factors

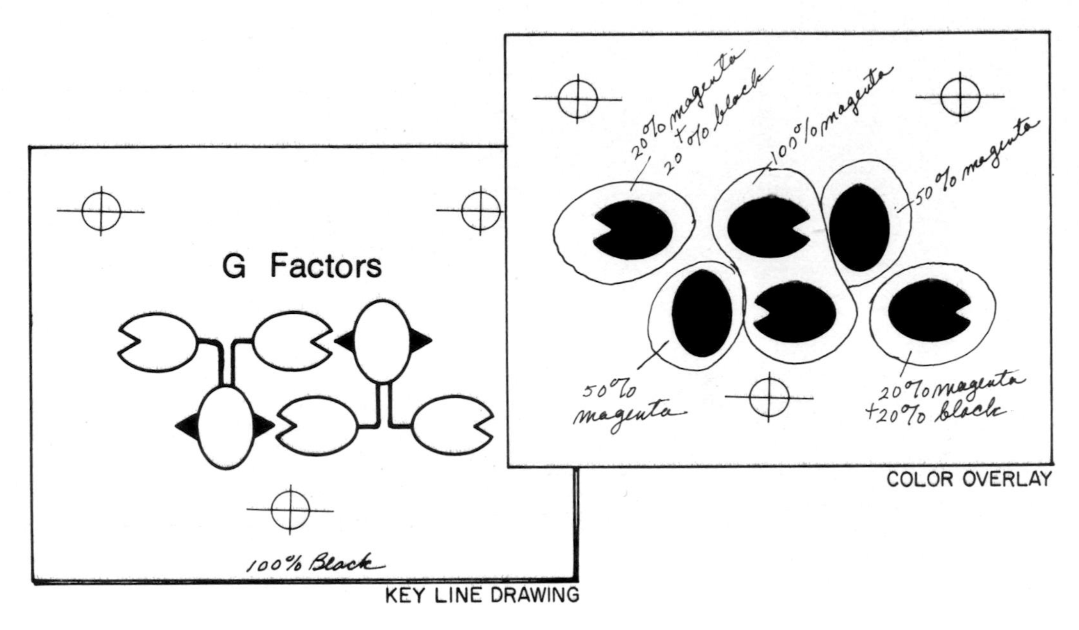
G Factors
100% Black
KEY LINE DRAWING
20% magenta + 20% black
100% magenta
50% magenta
50% magenta
20% magenta +20% black
COLOR OVERLAY

6-2.

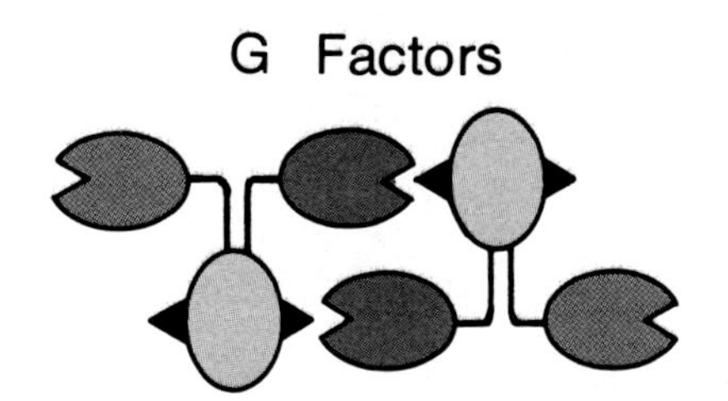
G Factors

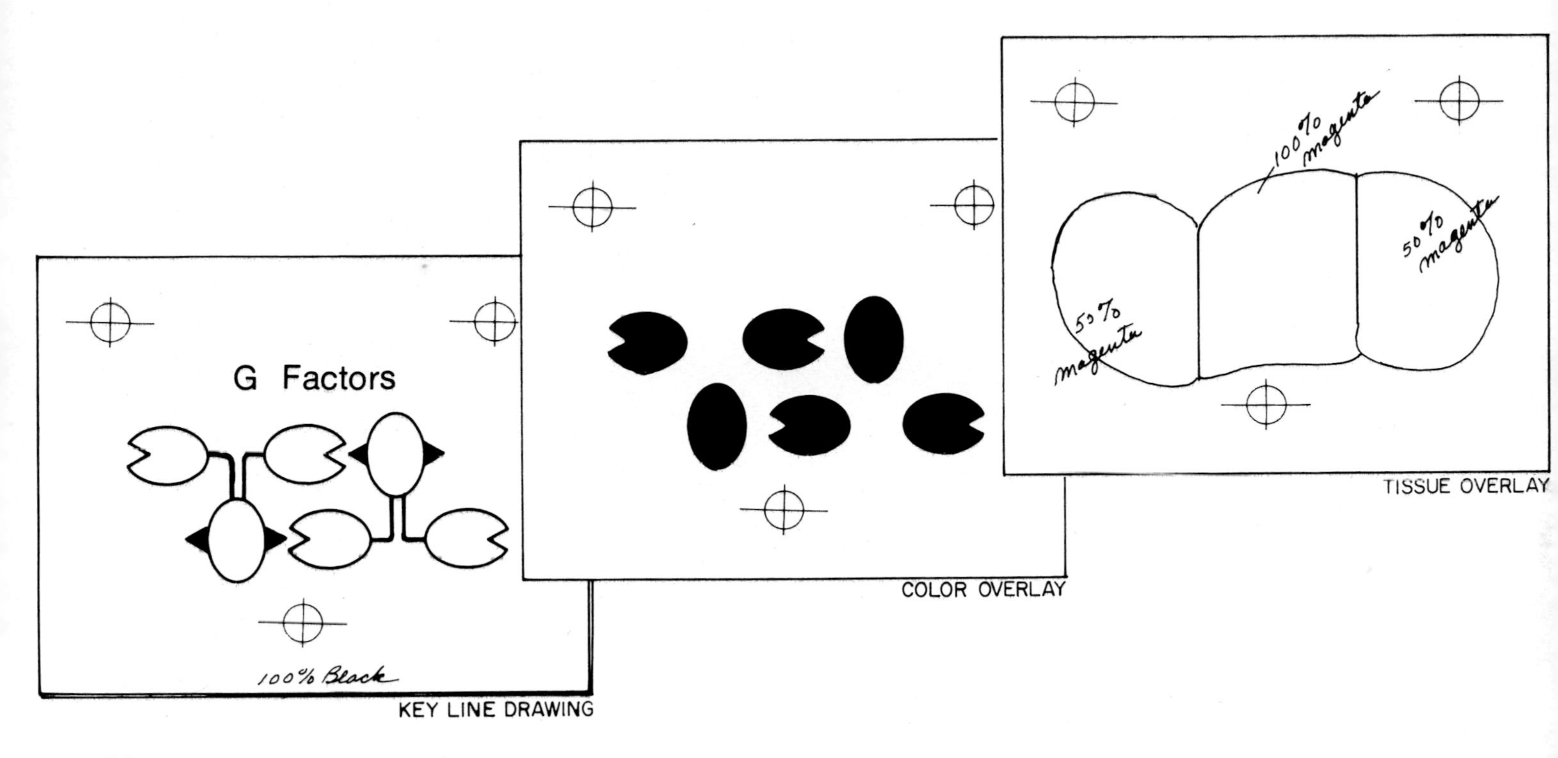
G Factors
100% Black
KEY LINE DRAWING
COLOR OVERLAY
100% magenta
50% magenta
50% magenta
TISSUE OVERLAY

6-3.

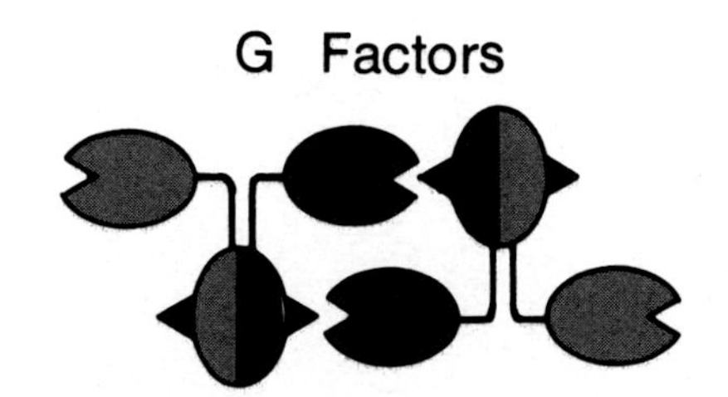
G Factors

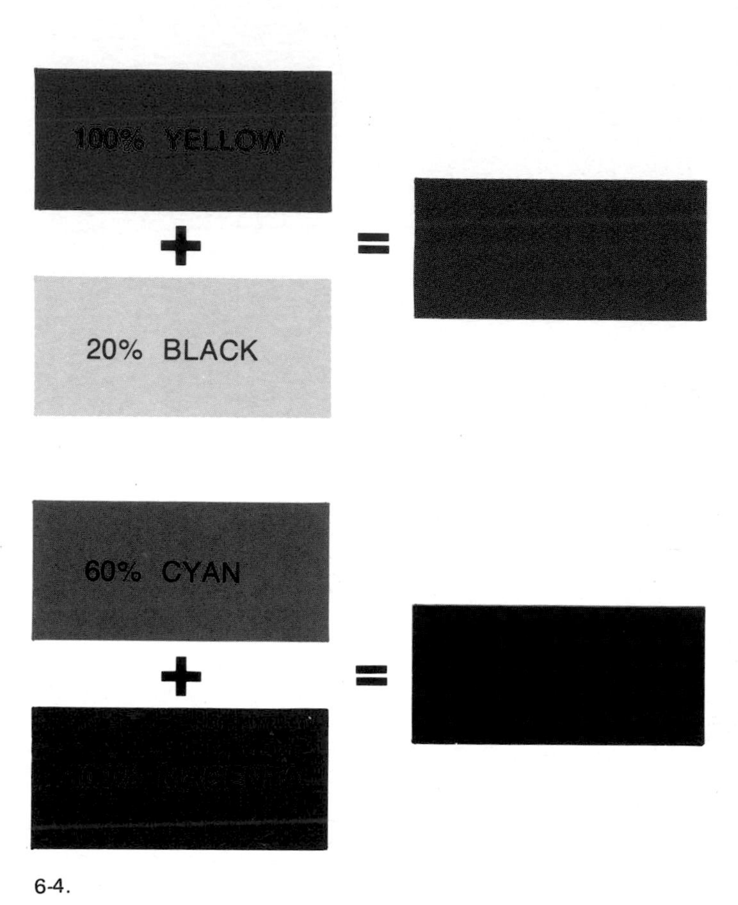

6-4.

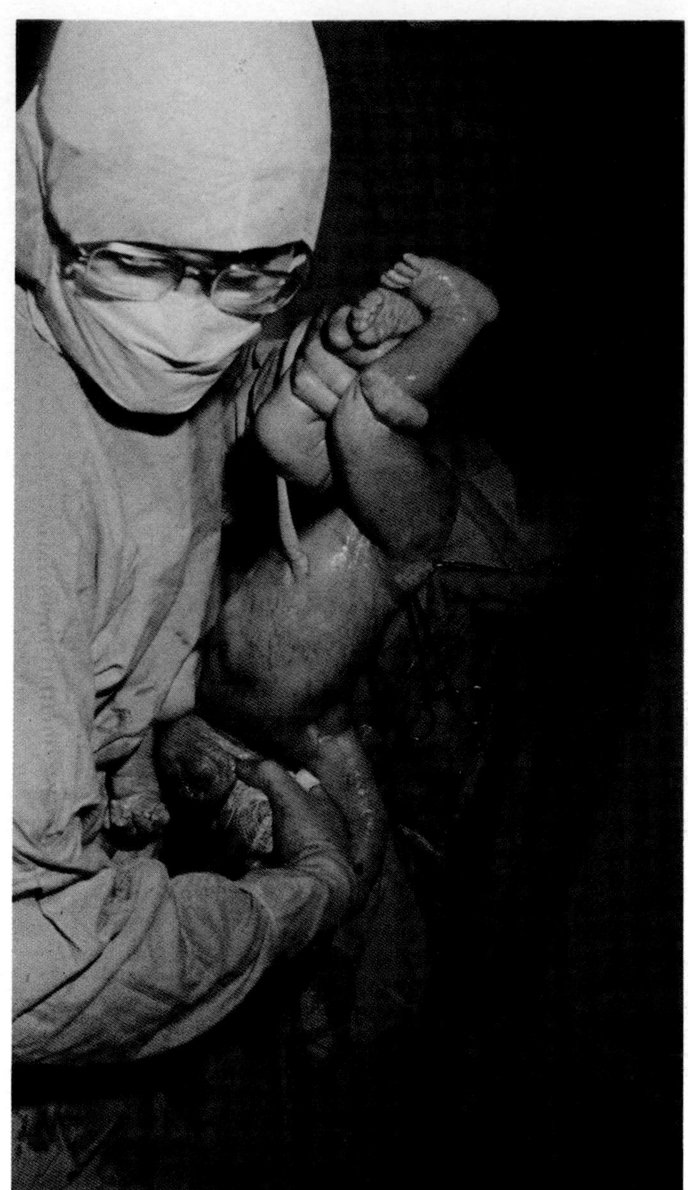

6-5. Marie Hanak.

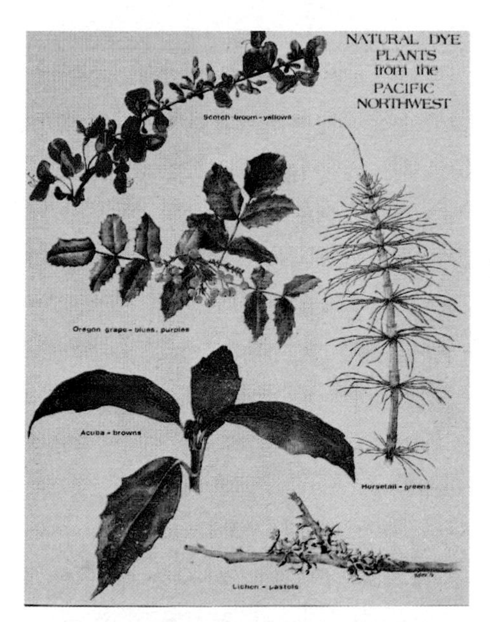

C-1. Margaret Watson. Watercolor and color pencil on cold-press illustration board.

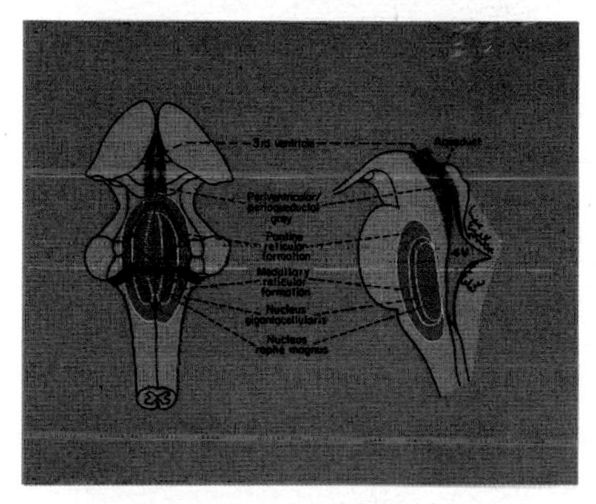

C-2. Phyllis Wood. Brain stem. Color-paper cutout with Transparex overlay.

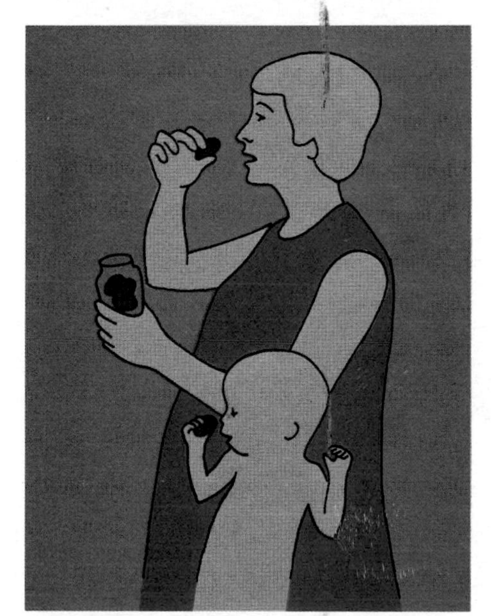

C-3. Phyllis Wood. Preventive pediatrics. Color-paper cutout with black Transparex overlay.

C-4. Elizabeth Keohane. Iris. Watercolor and color pencil on cold-press illustration board.

C-5. Yellow printer.

C-6. Cyan printer.

C-7. Magenta printer.

C-8. Black printer.

C-9. Rosettes.

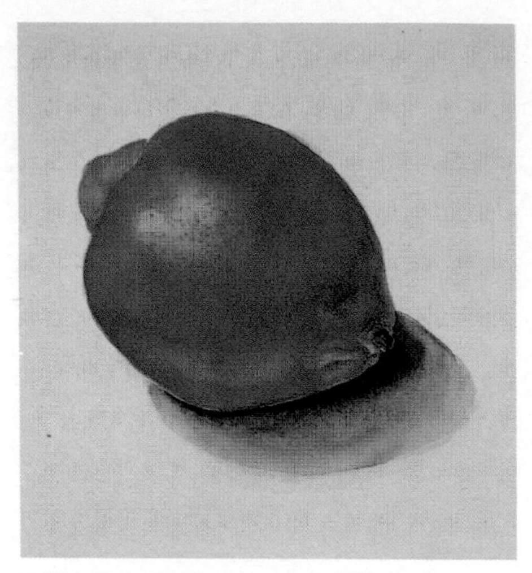

C-10. Marjorie Davis. Lemon. Watercolor on cold-press illustration board.

C-11. Cheryl Vigna. Tomato. Watercolor on cold-press illustration board.

C-12. Marjorie Davis. Cabbage. Color pencil on cold-press illustration board.

C-13. Nancy Williams. Watercolor on cold-press illustration board.

C-14. Phyllis Wood. Corn. Color pencil on paper.

C-15. Jane Rady. *Physalis* (Chinese lantern) seed pod. Watercolor on cold-press illustration board.

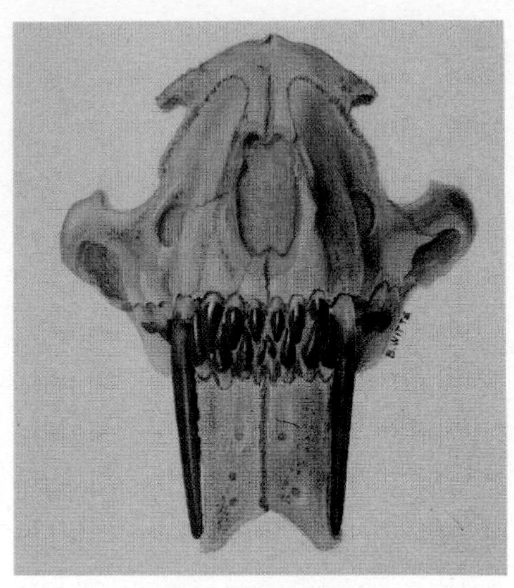

C-16. Beverly Witte. *Hoplophoneus primaevus.* Watercolor and pencil on cold-press illustration board.

C-17. Kathleen Todd. Tomato. Color pencil on video paper.

C-18. Kathleen Todd. Rhododendron. Color pencil on video paper.

C-19. Beverly Witte. Navel orange. Watercolor on cold-press illustration board.

C-22. Marcy Gordon. Strawberry. Watercolor on cold-press illustration board.

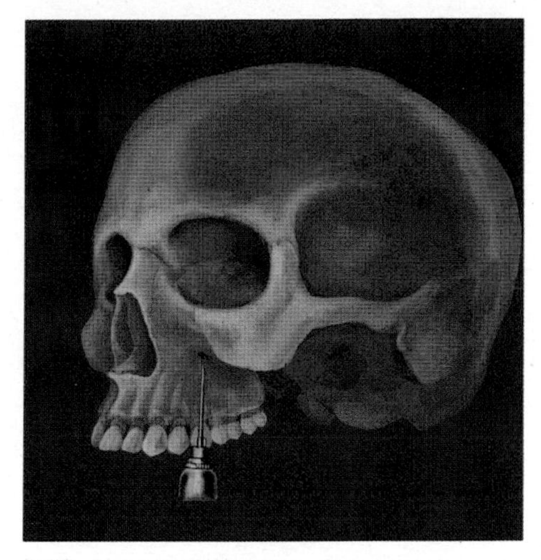

C-25. Nelva B. Richardson. Nerve block. Acrylic on cold-press illustration board.

C-20. Dinah Wilson Stone. Onion. Watercolor on cold-press illustration board.

C-23. Shirley Reiss. ***Hyalophora cecropia*** male. Watercolor and pencil on cold-press illustration board.

C-26. Nöelle Congdon. Color pencil on video paper.

C-21. Margaret Watson. Jerusalem artichoke. Watercolor on cold-press illustration board.

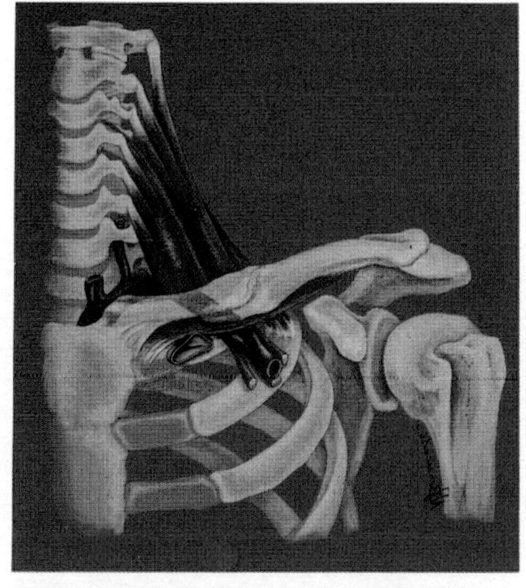

C-24. Nelva B. Richardson. Broken clavicle. Acrylic on cold-press illustration board.

C-27. Beverly Witte. Color pencil on black cold-press illustration board.

C-28. Joel Ito. *Galago crassicaudatus* (bush-baby). Airbrush on cold-press illustration board.

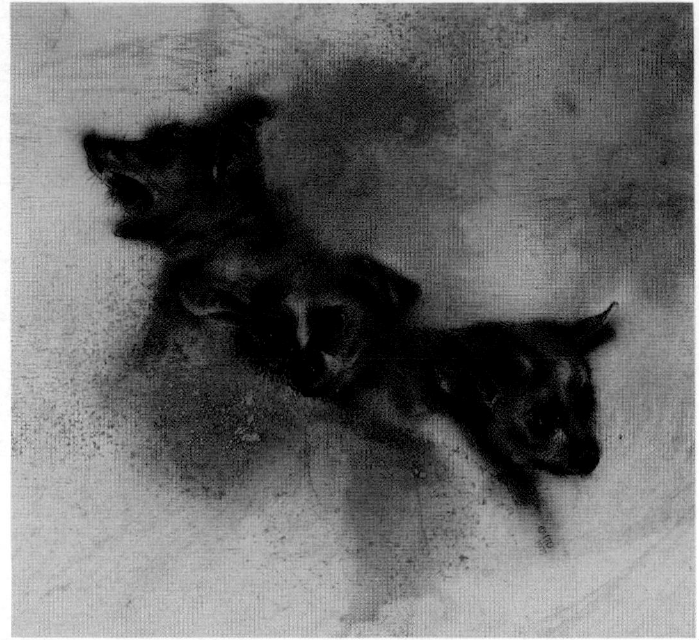

C-29. Joel Ito. *Galago crassicaudatus* (bush-baby). Airbrush on backpainted frosted acetate.

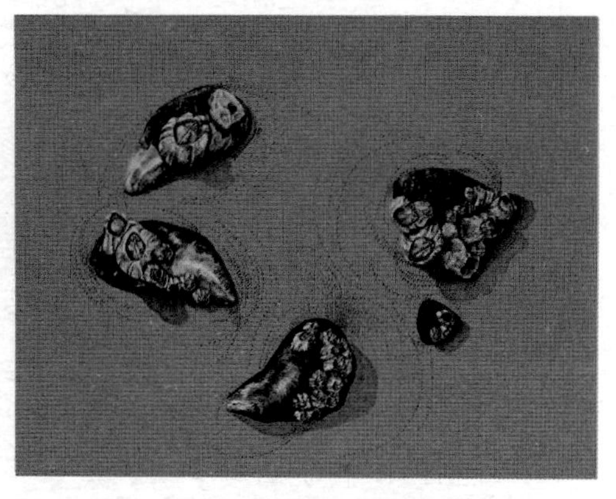

C-30. Nancy Williams. Cockles. Watercolor, color pencil, and ink on charcoal paper.

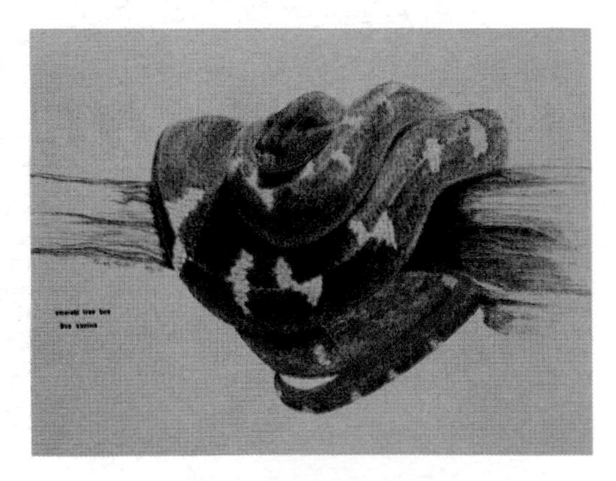

C-31. Pamela Harlow. Color pencil on video paper.

C-32. Kathleen Todd. Arab stallion. Color pencil on video paper.

C-33. Kathleen Todd. Arab stallion. Color pencil on video paper.

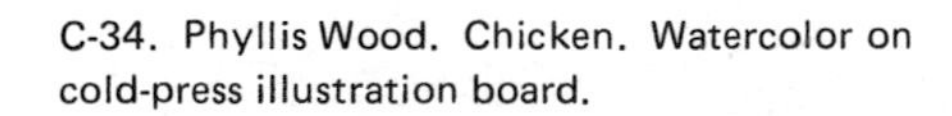

C-34. Phyllis Wood. Chicken. Watercolor on cold-press illustration board.

CHAPTER 7.

Animal Illustration

In the chapter on drawing (Chapter 2) emphasis was placed on familiarity with the subject, acquaintance with the structure and function of its parts, and understanding of its classification within its particular discipline. In no area is this more important than in that of the live animal, and in no area is it more difficult. You cannot sit at a drawing board and carefully measure and draw the specimen, as you can a stationary subject. The source material for the illustration may be a live and moving animal, a photograph, a preserved specimen, or merely a skeleton, but the end result must have all the characteristics of the live animal.

7-1.

THE LIVE ANIMAL

A sketch pad, pencils (color, carbon, or graphite), a pencil sharpener, a folding stool, and perhaps binoculars are all you need for the first phase of the illustration. You should use color pencils or watercolors if the final drawing is to be done in color.

Character

Arrange a quiet and comfortable place as close to the animal as possible. Plan to sit quietly in one position for a period of time, at least an hour, while you make attitude sketches. During this period the animal will hopefully become accustomed to your presence and ignore you, assuming characteristic positions. The most important attitudes for the artist to practice at this stage are patience and persistence (7-1).

The first sketches should not be serious but rather loose and quickly done to capture the essence of the animal. As you do many of these free character impressions, you will begin to get the feel of drawing this specific animal and to develop a rapport (7-2).

It may be a frustrating experience, as the animal may decide to sleep through it or to move so quickly that it is difficult to capture characteristic attitudes on paper. Patience, persistence, and practice are useful. You may decide that a sitting at a different time of day would be more fruitful. Find out about the animal's daily cycle of eating, sleeping, and activity and about changes in disposition.

After this preliminary get-acquainted period you can decide on the stance that will suit your purpose best, showing the animal most advantageously and in a natural position (7-3).

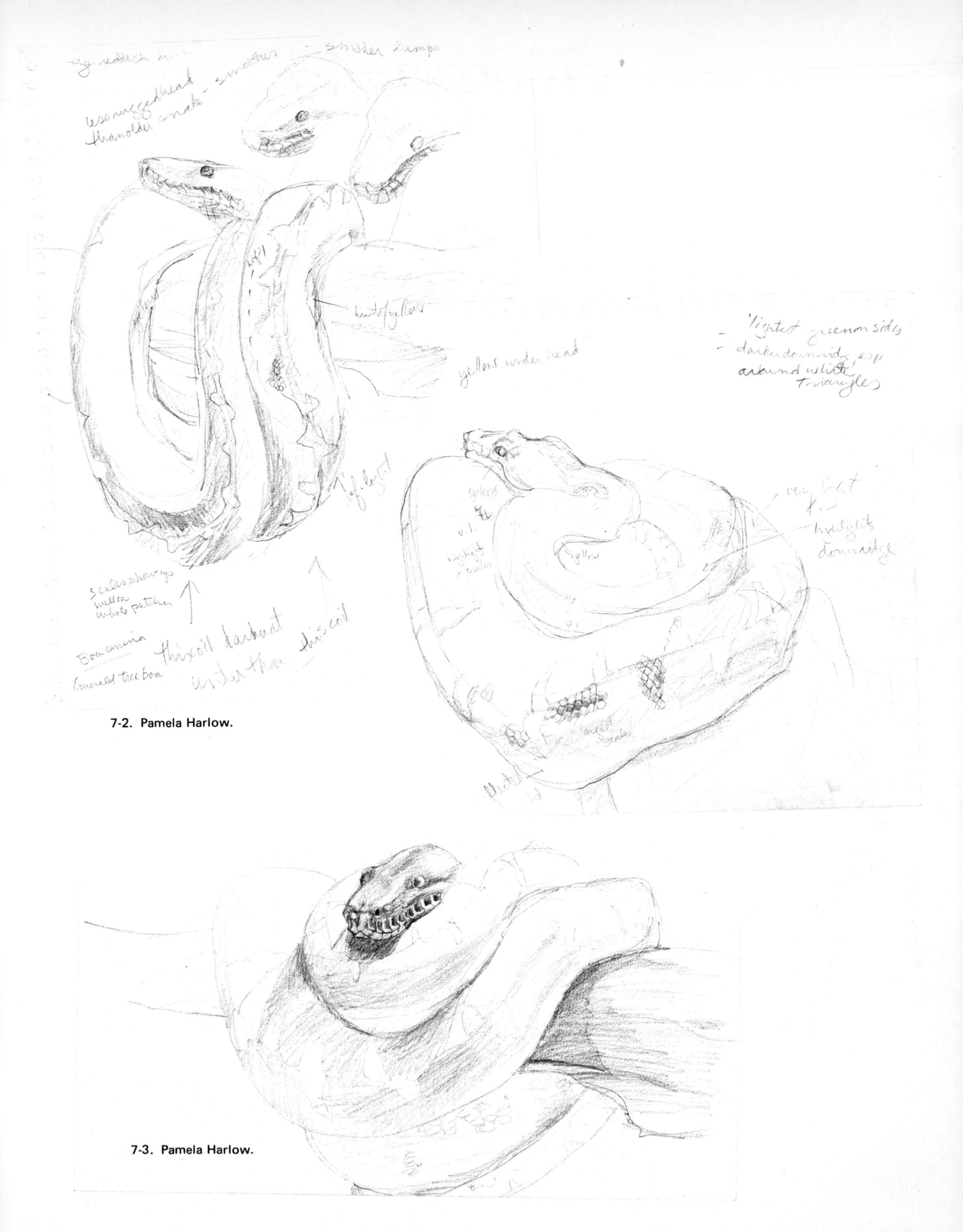

7-2. Pamela Harlow.

7-3. Pamela Harlow.

Details

After you have decided on and drawn the general posture and proportions with some finality, study and draw the details carefully. The following is a brief sample of the various elements that must be studied critically.

1. Fur: changing directions and length, direction at the joints, smoothness or roughness
2. Face wrinkles and brow hairs
3. Mosaic patterns: regularities and variations
4. Eyes: lid shadow, highlight, secondary highlight at the opposite pole of the orb
5. Legs and feet: characteristic bending in relation to each other, relation of nails and toes
6. Color: highlight areas, middle-value areas, shadow areas, effect of surface hair on crispness of color pattern

These features should be worked in tight detail studies separate from the primary drawing.

Final drawing

Take these sketches back to the studio to form a basis for the final drawings. Remembering that every structure is an element in the interrelationship of the whole animal, assemble the separate drawings in a natural and lifelike way. They should be integrated with the spontaneous quality of your attitude sketches.

The more you know the animal and are aware of the underlying structure of bone and muscle that is the framework for the visible surface, the more lifelike and successful your drawing will be. A scientist who is familiar with the animal can often make a more convincing drawing than an artist who is only aware of the surface.

As you develop a tight rendering from your sketches, you will probably find that what you thought were complete reference drawings leave some aspects of the animal's features still in doubt. You will nearly always find that several drawing sessions are necessary to complete a study satisfactorily (C-31).

REPRESENTATIVE AND INDIVIDUAL SUBJECTS

Each animal is as unique in his physical and emotional characteristics as is a human, so care must be taken to observe several members of the same species in order to draw an anonymous generalized animal that has the combined features of several specimens. Drawing a specific animal as a representative of an entire species would be as fallacious as using a picture of Rosa Bonheur as the prototype of *Homo sapiens.* Differences in age, sex, origin, and cyclic changes must be taken into consideration. The essence of these instructions is to know what you want to draw: a representative animal, a specific individual animal, or a representative of a group (adult male, juvenile female, infant, female in estrus, male in courting plumage, or rare subspecies).

To illustrate an animal properly, several drawings are often required. A group combining an adult male, an adult female, and a juvenile may be necessary to illustrate the species fully. The individuals should be positioned in such a way that their differences are obvious. It is helpful to show them in different views so that the same features are not repeated and the maximum amount of information is shown in the least amount of space (C-32).

Some animal drawings require much less information. Variations in color or body shape among species may be all that the assignment requires. A simple ink outline may be all that is necessary. Do not draw every hair or scale unless it is essential. Saturating the viewer with more information than the illustration requires only confuses rather than informs (7-4).

PHOTOGRAPHY

Even the strictest purist no longer feels that the use of photography corrupts his integrity. Polaroid film, high-speed photographs, and motion pictures are all used as an aid to stop-action and to confirm and document various proportions and postures of the subject. The demoralization of the art occurs when an artist who is not on speaking terms with the subject attempts to illustrate it using static photography as his sole source of information. Parallax, inherent in photography unless corrected, will be apparent. Definition of related structures and details may be ambivalent. The flash of the camera may flatten contours, obliterating the shadows that define form and creating deep shadows that obscure it. Photographs, although they do not replace the sketchbook, are a valuable auxiliary reinforcement to the expeditious development of the animal drawing (3-32).

THE PRESERVED SPECIMEN

Anthropological museums have collections of preserved specimens, but relatively few of these are reconstructed in a lifelike position. Most are preserved as rolled skins, with the insides removed and stuffed lightly to prevent flattening. They can, however, be an abundant source of information. Details of hair and feather arrangement, exact proportions, structure of the extremities and the face, and color patterns can be studied closely. Stuffed specimens placed in lifelike positions can be observed from all sides, showing the interrelationship of parts and the foreshortening in perspective.

You should keep in mind that all bone and tissue are removed from the stuffed specimen, leaving only the thin skin with its outer covering of scales, fur, or feathers. A technician, scientist, or taxidermist with much or little experience in reconstruction and varying degrees of knowledge about the specific animal has attempted to restore it to its original shape and size in a natural position with solid, stiff materials unlike the animal's own understructure. It is easy to see how some unnatural features may creep in during this very demanding and delicate procedure. The stuffing may not conform to the actual shape or size of the living animal. The feet and legs may be in an unnatural position or even reversed. The glass of the eye may be more creative than authentic. Some parts (bills, feet, etc.) must be painted to approximate the colors of the living animal. These colors may be difficult or impossible to match and may have been inadequately researched. In any case preserved specimens should be used as a reference only in combination with other resources and not as the sole source of information.

7-4

Pan troglodytes
Chimpanzee

7-5. Phyllis Wood. Ink on scratchboard.

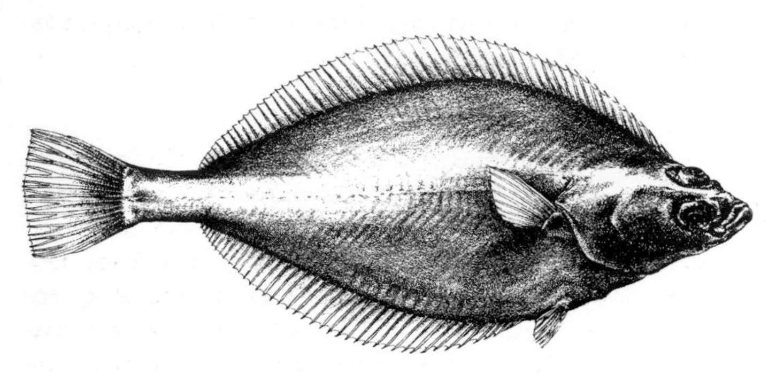

7-6. Anton Friss. Fish. Litho pencil on coquille board.

TECHNIQUES

When contemplating the technique to use for an animal drawing, the first and most natural choice is color, because the empathy that we feel for animals makes it hard for us to separate them from their color. Any of the color media (watercolor, acrylics, airbrush, and even color pencils) are appropriate. Continuous tone using wash, carbon dust, pencil, or airbrush can be effective. In black-and-white drawing ink line and stipple have different advantages.

The best way to choose a technique within the confines of ultimate use (print, film, etc.) is to study the surface of the animal. A wet or shiny surface is best suited to wash, watercolor, or color pencil (C-33). A hairy surface should be interpreted in a technique that suggests hair—line drawing, particularly on scratchboard (7-5), or drybrush over a smooth underpainting. Smoothly contoured surfaces are well rendered with stipple, coquille board (7-6), or any of the smooth techniques—wash, airbrush, or carbon dust. Pencil and watercolor suggest the immediacy of attitude sketches, although both can also be worked up smoothly and with great detail (7-7, C-34). A thoughtful look at the specimen coupled with a knowledge of printing requirements can help you make an informed decision on technique.

BACKGROUND

Most animals, if drawn accurately in their natural environment, would be camouflaged so effectively by the light, shading, and surroundings that they would be difficult to see. As the purpose of most illustrations is to clarify the subject, the background and the atmospheric impression must be deemphasized in favor of subject clarity.

7-7. Phyllis Wood. Giraffe. Color pencil on ledger paper.

7-8. Dean Rocky Barrick. Dipper. Graphite pencil on museum board.

7-9. Marjorie McKinley. Lion. Wash on cold-press illustration board.

Simple complementary backgrounds, however, add a great deal to the information and charm of animal drawings. A branch against the sky, some leaves on the ground, or water on the beach helps to place the animal in the rightful habitat and in the proper scale. Care must be taken to use background materials from the animal's native habitat, which are not necessarily the same as those found in an adopted environment (7-8).

DOCUMENTATION

Each sketching session should be documented with field notes, including the scientific and common name of the animal, its sex and age, the date and time of day, and the observation site. If you are drawing the animal in its natural environment, be specific about the area and its description. Notes should be made of the animal's actions, both generally and specifically and alone and interacting with other animals. If the specimen is not alive, records should be kept of the condition: whether freshly killed and where, type of preserved specimen (stuffed, rolled), and exact location of the collection. This documentation, while interesting in itself, also proves invaluable as reference material.

LIFELIKENESS

The key to animal drawing is lifelikeness. The satisfaction of completing an animal drawing that mirrors the attitude and unique character of that particular animal is worth all the patience, persistence, and practice that it takes to achieve. Use all the resources that are available to you: photographs, motion pictures, slides, fresh and preserved specimens, and firsthand observation. Avoid recreating the stiff solidity of the stuffed specimen or the flat and static character of the photograph. Remember that there is a living, moving, breathing animal in your drawing and stretch yourself emotionally and artistically to infuse life into your creation. Try for the unusual angle, the distinctive gesture, the exact detail, the spare and personal background. Nothing in scientific illustration is more rewarding than an animal drawing with the fresh, spontaneous quality of the animal itself (7-9).

CHAPTER 8.

Diagrams

In a research facility the artist-chartist can expect that up to 75% of his drawings will be diagrammatic artwork, including statistical data, schematic or abstract ideas, and maps. The artist should understand the information and be able to visualize and to produce it in graphic form using cost-effective methods. While the effectiveness of the message is of prime importance, the diagram should be designed as thoughtfully as any other piece of artwork.

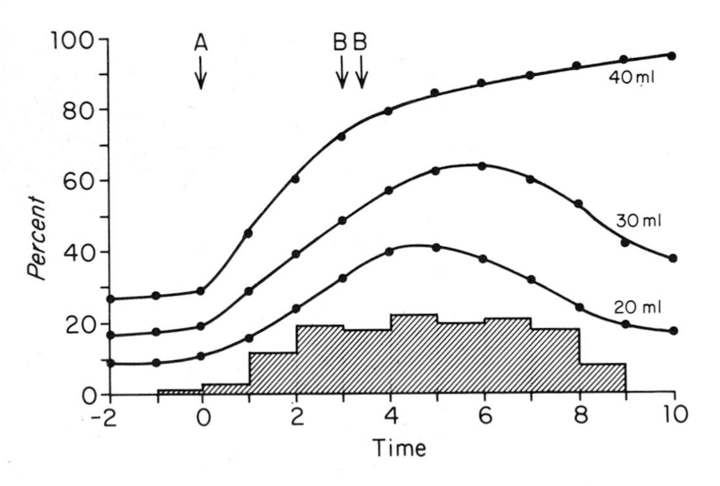

8-1.

PURPOSE

A diagram is used to clarify and reinforce a verbal or written presentation. It should be coordinated with the verbal presentation (slide lecture, television, motion picture) or with the written presentation (book or publication).

A diagram can illustrate statistical data that would be very ponderous to explain with tables of figures (8-1). It can contemplate abstract concepts (8-2). It can show interrelationships immediately (8-3). It can telescope or build up complex interactions in a series (8-4). In order to accomplish all these miracles, it must convey a spare amount of information and be straightforward in design so that it can be absorbed quickly.

While the information presented is serious, factual, and scientific, there is no reason for the end result to be dull. Originality and even occasional levity increase the retentive quality of graphs, schematics, maps, and statistical data (8-5).

DESIGN

Simplicity of design and singularity of idea are important. A design should stress the important thrust of the message, with less emphasis on the supporting data. The decisive question in planning a design and the amount of information to include is whether there is an immediate impression of the diagram's intended message.

Bright colors, bold shapes, and heavy lines are used for the important data. They tell the viewer where to look. Neutral colors, subordinate shapes, and light lines are used for the supporting information.

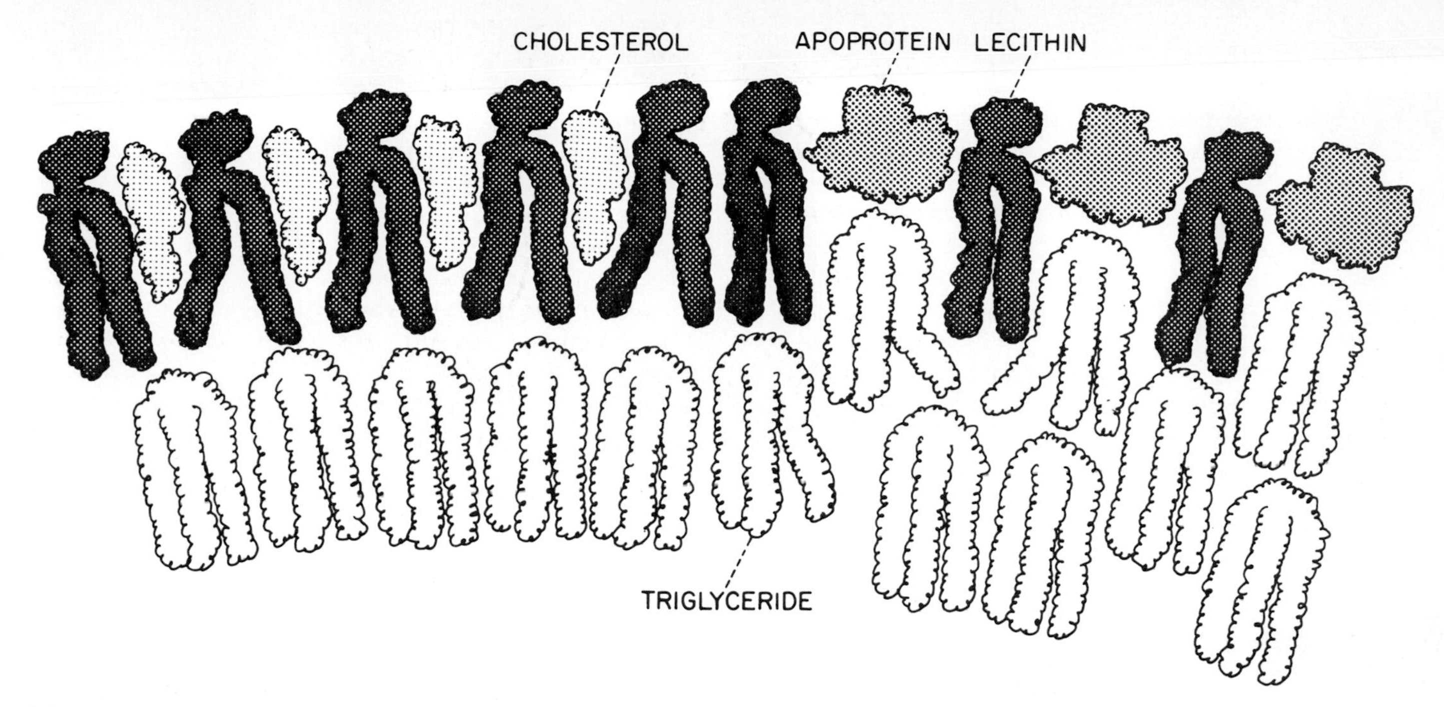

8-2.

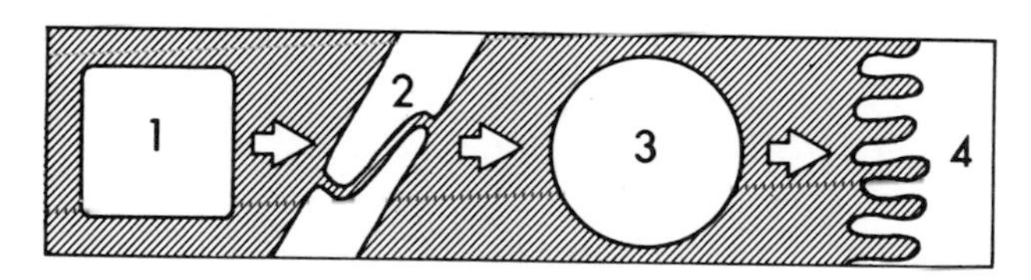

8-3.

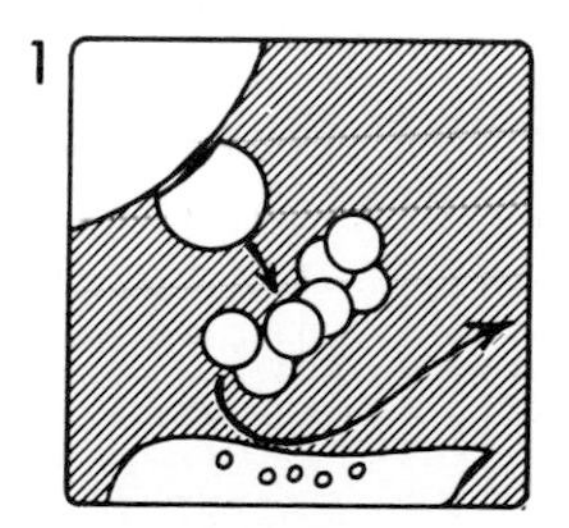

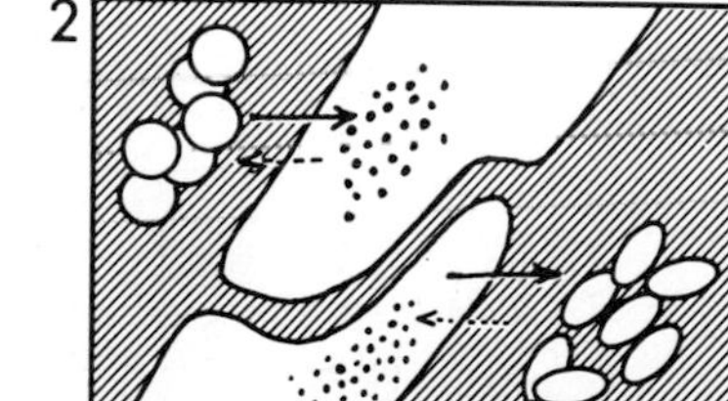

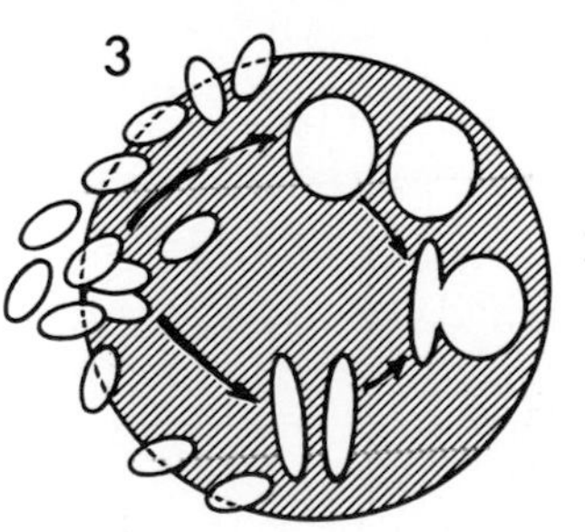

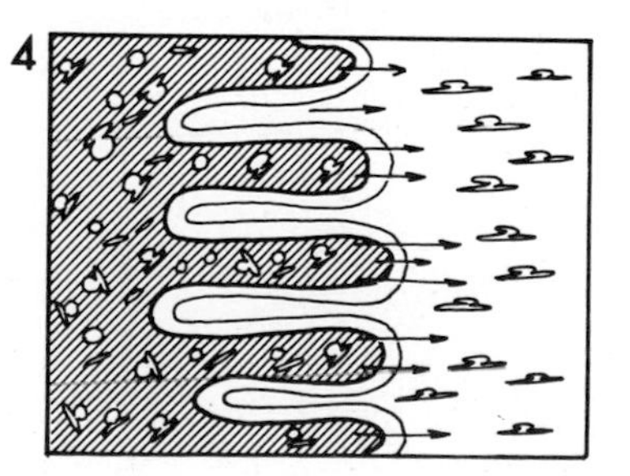

8-4.

1. THE HEART AND CIRCULATORY SYSTEM

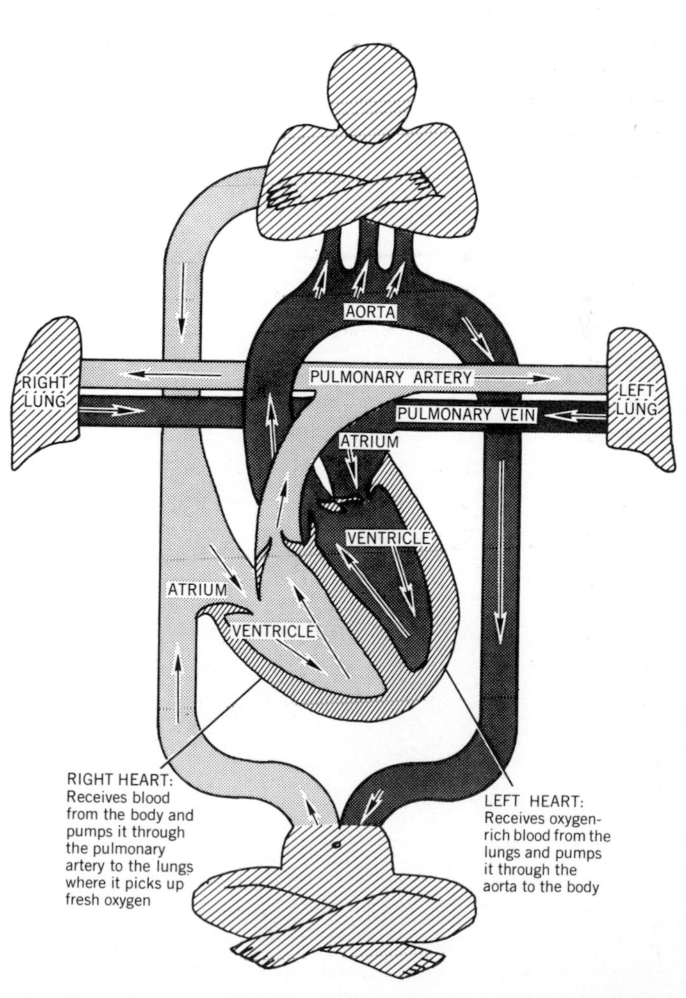

8-5. From Donald W. Miller, Jr. *Bypass Surgery for Coronary Heart Disease*, 1977.

PRINT MEDIA

The importance of eye-catching and easily understood diagrams in print media becomes more evident when one is aware of the great proliferation of written material competing for the reader's attention. The reader has time merely to scan most publications, looking for the subjects that capture his interest. He often looks only at the diagrams. He may be satisfied by reading the captions, or they may perhaps stimulate him to read the entire article. Diagrams have a great responsibility to a publication.

While it is possible to include a good deal of statistical data in a printed page, it is usually better to break material into smaller units of data than to overwhelm the reader with too complex an idea. More detailed information may be included in a printed diagram than in a projected one, however, because the reader can study the data at his own pace.

Scaling

Diagrams are drawn larger than the printed size. The reduction percentage to be used should be carefully considered when selecting the size of the lettering and the width of the lines. The lettering should be similar in size to the body copy of the publication. Lettering that is much smaller than the body copy is difficult to see and probably won't be read. Lettering that is much larger will be gross and ugly.

The diagram should be planned to fit a specific increment of the page layout. The intended page size, column width, and general design policy should be observed. The diagram can be planned as a strip across the width of the page, fit into a vertical column, or scaled to fill any part of these areas or a complete page (see 9-28). This decision is made according to the amount of detail in and the proportions of the diagram. The more the detail in the diagram, the greater the space needed so that it is comfortable for the viewer to read.

Black and white or color

Diagrams are generally printed in black and white. Either the original or a good copy can be sent to the publisher. In a good copy every black area is clear and decisive, not gray or interrupted, as grays *do not* print. The white areas must be clean and unsmudged, as smudges *do* print. Examine photocopies carefully for irregularities before sending them. If copies are sent, they should be either the same size or larger than the printed size. Color diagrams are prepared in flat color, preseparated by the artist on overlays (see 6-7, 6-8, 6-9).

NONPRINT MEDIA

The most important aspect to remember in designing graphics for projection is that they should complement and not compete with the oral communication (slide lecture, television, or motion picture). The viewer should listen to the lecture and not read it on the screen. The graphic should make its point quickly with a minimum of words. Each slide should emphasize only one point. A complex idea may need to be broken into several separate slides (8-4). The confusion factor rises sharply with the use of projected rather than printed graphics, so simplicity is of prime importance.

Proportion

The diagram should be designed to utilize the entire projected area. If the proportions vary too much from this, the material will be reduced more than is necessary (8-6).

Lettering should be done with consideration for the audience. Large, bold lettering is necessary in order to be seen at the back of a large auditorium (8-7). Slides to be projected in a small conference room need not be quite so substantial.

Thin lines or delicate detail may be difficult to see or may disappear completely when projected. The guideline for all projected diagrams that is **BOLD IS BEST.**

A good general rule is to include no more than seven lines of lettering on a slide and no more than seven words across a slide. The height of the lettering for television should be at a minimum 1/14 the height of the screen. The important data should be near the center, as the periphery is not dependably readable.

Color

Color is easy to produce and relatively inexpensive in nonprint media. The dramatic qualities of color aid memory if used in a thoughtful manner. Continuing symbolic color through a series of visuals is helpful (8-8). Emphasis and deemphasis are valid uses of color. Merely to make a slide bright is not a valid use of color, and it may confuse rather than inform the viewer. Contrasting colors on the original may not contrast well when projected, so pretest them before launching a project.

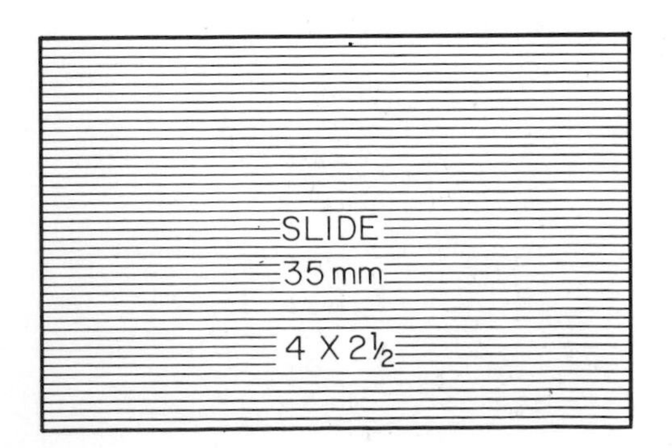

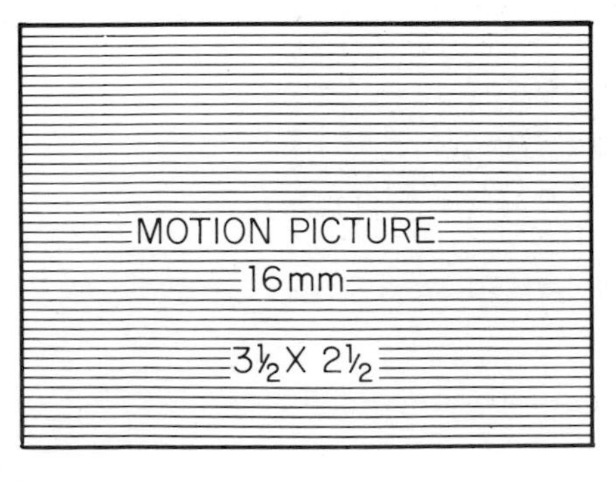

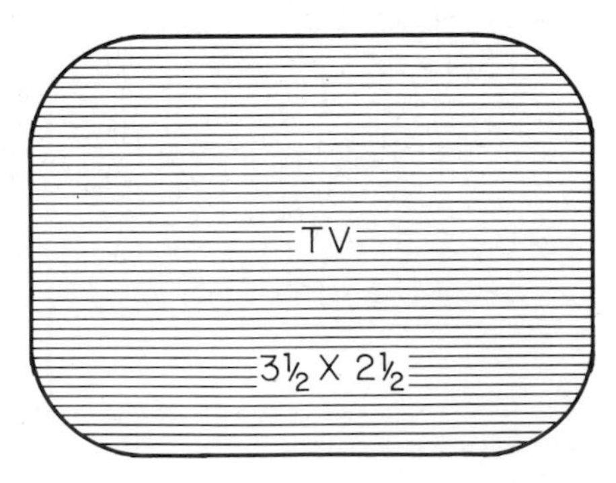

8-6.

BOLD
LETTERING
PROJECTS
BEST

8-7.

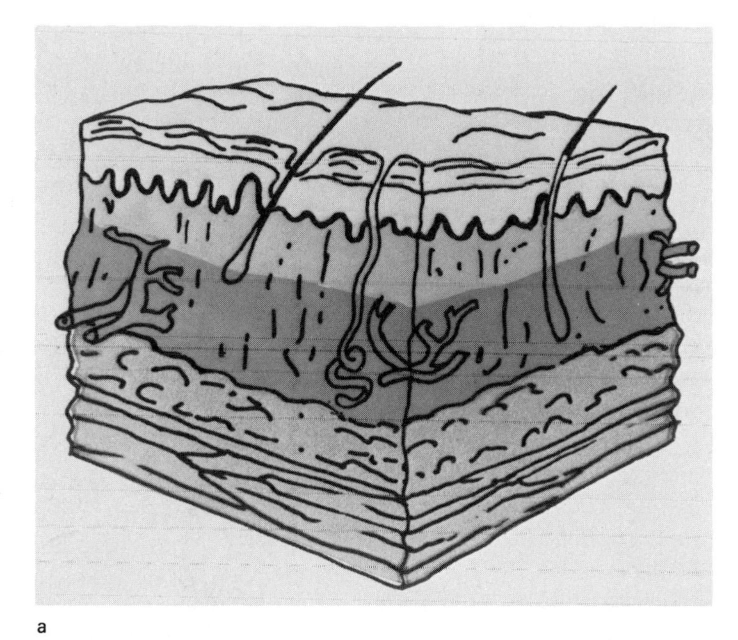

a

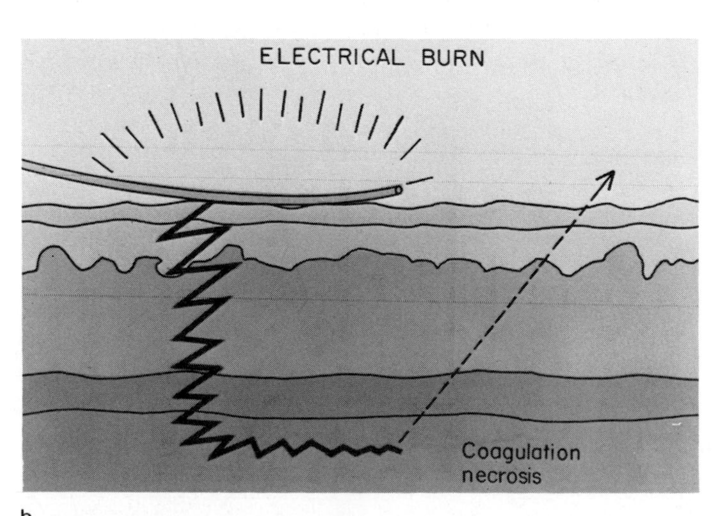

b

8-8. a. Nancy Todd Nelson. b. Nancy Todd Nelson. Black Transparex film over color paper.

MATERIALS

A diagram is either traced from the pencil layout onto translucent paper or film or transferred onto opaque paper or board. Mechanical pens and ink are used with templates to trace the diagram, or purchased tapes, symbols, and areas of film or paper are affixed.

Grounds

As there is such a wide range in type and quality among various translucent papers and films, they should be purchased carefully. Before stocking any new paper or film test it for smooth ink acceptance, erasure without surface marring, and crisp scratching to remove small areas of ink.

Translucent paper should be heavy two-ply with an even translucence and no visible hairs or flecks. The surface should be hard so that the ink flows evenly on it. Films (Cronaflex, Mylar or mat acetate) should be mat-finish on at least one side, which is the side that you draw on. They should be free of imperfections.

Most films wash clean with water and a cotton swab. Papers can be erased with an electric eraser and a white pencil-eraser insertion. Ink can be scratched from both types of surface with a blade. When working with black-and-white copy, areas can be spliced together or excised without showing on the final copy.

Films and papers come in rolls, single sheets, and pads. The pads vary from 8 1/2″ X 11″ (21.5 X 28 cm) to 19″ X 24″ (49 X 60 cm). Illustration board provides a sturdy mounting for color diagrams and comes in a good variety of background colors. It has a subtle texture that photographs well for slides. It comes in sizes up to 32″ X 40″ (80 X 100 cm).

Pens

Mechanical pens are used because they maintain a constant line width in a wide variety of sizes. The Leroy system consists of a standard pen holder or a scriber into which a pen point can be inserted. The pen point consists of a bucket to hold the ink and a cleaner pin that releases the ink when it contacts the paper. No pressure is necessary to initiate the ink flow (8-9). The bucket and pin are separated, washed, and blown dry after every use. Opaque india ink is used. Pelikan T is a good brand. Leroy pens can also be used with colored inks, Dr. Ph. Martin's Radiant concentrated watercolors, or dilute poster paints. Color inks are difficult to apply evenly, however, and the colors are usually not vivid.

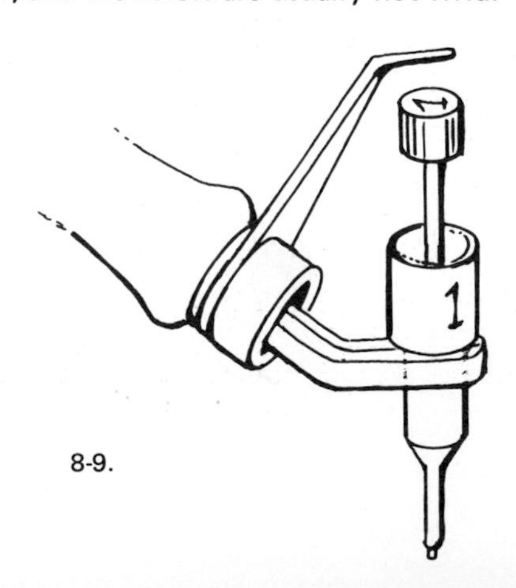

8-9.

In technical fountain pens with ink reservoirs the pen and the point are attached: examples are Castell TG and Rapidograph (8-10). Great care must be taken to prevent the ink from drying in them. They should be taken apart carefully and washed and dried thoroughly when refilling with ink. They come in a large range of sizes. The TG pen has a moist indicator cap. These pens use black ink made especially for them.

Felt-tip pens are quick and can be used for informal presentations on white or light backgrounds. They come in many colors and widths.

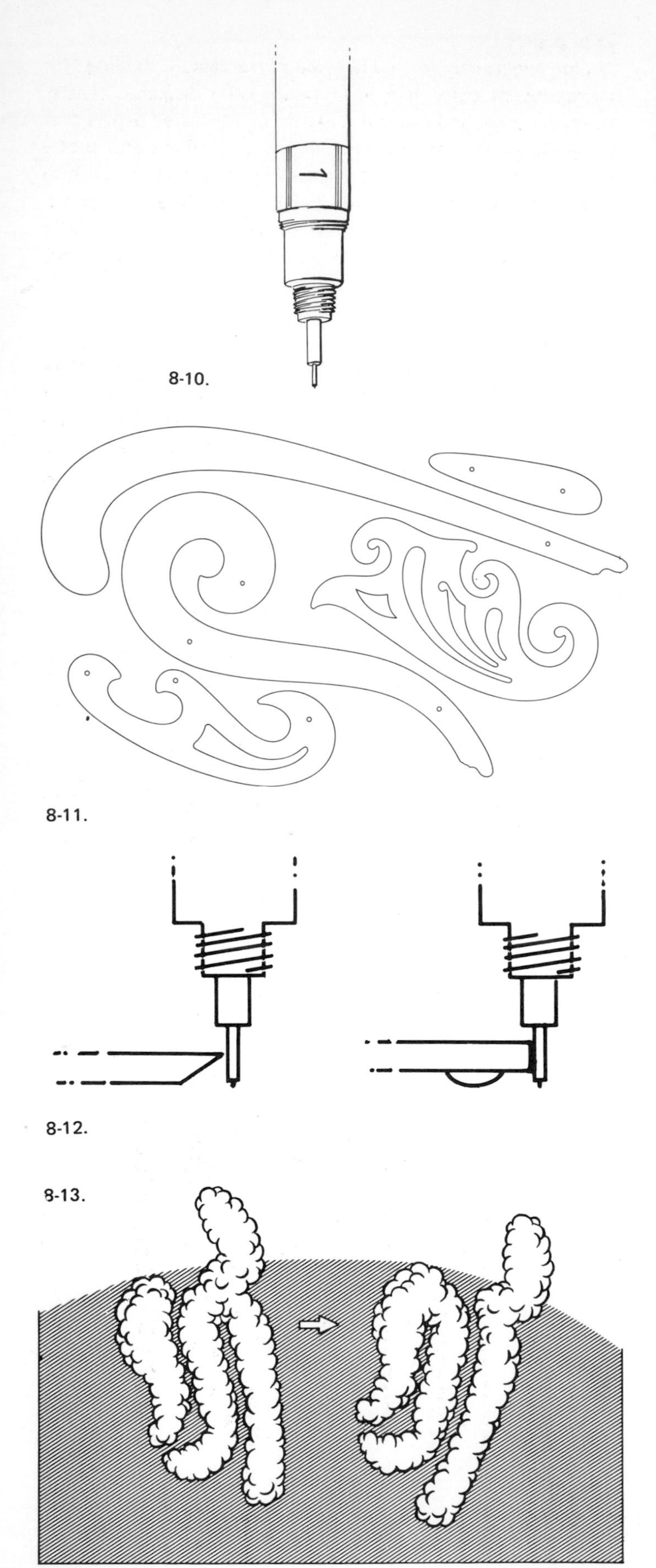
8-10.

8-11.

8-12.

8-13.

a

Templates and curves

Plastic templates are used to produce smooth curves and geometrics that are consistent in size and shape. Templates come in many sizes and degrees of ellipses, circles, squares, triangles, hexagons, arrows, and other shapes. There are two general kinds of curves. French curves include varieties of long, tight, and reverse curves. Ship's curves are long, gradual, and one-directional: they describe the curves used in designing a ship (8-11).

When tracing the template, hold the mechanical pen in a vertical position. If the template is not beveled, take care that the ink does not bleed under it. Single drops of nail polish applied to the surface of the template will help to hold it off the paper, substituting for a bevel (8-12). You may also raise the template off the paper with squares of tape or pieces of plastic. Leroy templates include a variety of symbols as well as lettering and are used with the Leroy scriber.

Shading films

Transparent shading films come in a great variety of patterns on adhesive-backed sheets. Some of the brand names are Chartpak, Letraset, Zipatone, and Formatt. They are useful for accenting an area or for differentiating between several areas and in general make the diagram more easily and quickly recognizable. They are sometimes used for the background to make the subjects stand out in *relief* and sometimes to *solidify* the subjects (8-13). Sheets of adhesive-backed translucent films are also available in many colors under the same brand names for color graphics. They are handled in the same way as are the shading films.

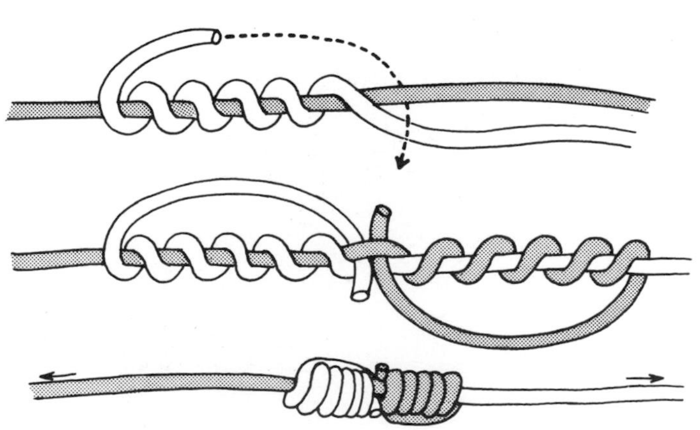
b

When several areas are differentiated with shading films, make sure that they appear as different values on the gray scale when reduced. While a stipple pattern may look quite different from a line pattern in actual size, they may appear to be the same value when reduced. Hold the shading films away from you and squint at them to see whether they separate as contrasting values on the gray scale (8-14). Very fine-line or stipple patterns may fade or disappear when they are reduced for publication or photographed for projection.

To use the shading film, place it adhesive side down on the area that you wish to cover, making sure that it is free of eraser crumbs and dust particles. Cut the film using a #11 scalpel blade or a #16 X-Acto knife. Use a light touch so that you cut through only the film and not the drawing paper. Remove the film from the unshaded area and rub it down in the shaded area with a plastic burnisher, your thumbnail, or any blunt, smooth instrument (8-15).

Color papers

Color papers can be used in place of transparent shading film or translucent color film in preparing color diagrams. Letrapaper, Pantone, and Color-Aid are thin, untextured papers that are available in a tremendous variety of brilliant and subtle colors. Cut the paper in the shape of the area that you wish to emphasize and attach it to the background. Small pieces can be attached with a glue stick. Larger pieces adhere more flatly if dry-mount tissue or Double Nothing is the bonding agent (see chapter 11).

Tapes and symbols

Adhesive-backed rolls of tape come in black and colors. They vary in width from 1/64″ (0.4 mm) to 2″ (5 cm). The black tape comes in many patterns. Mat-finish is best for photographic reproduction. Symbols come in rolls or sheets in black, white, and colors. They include geometric shapes, arrows, borders, and stars (8-16).

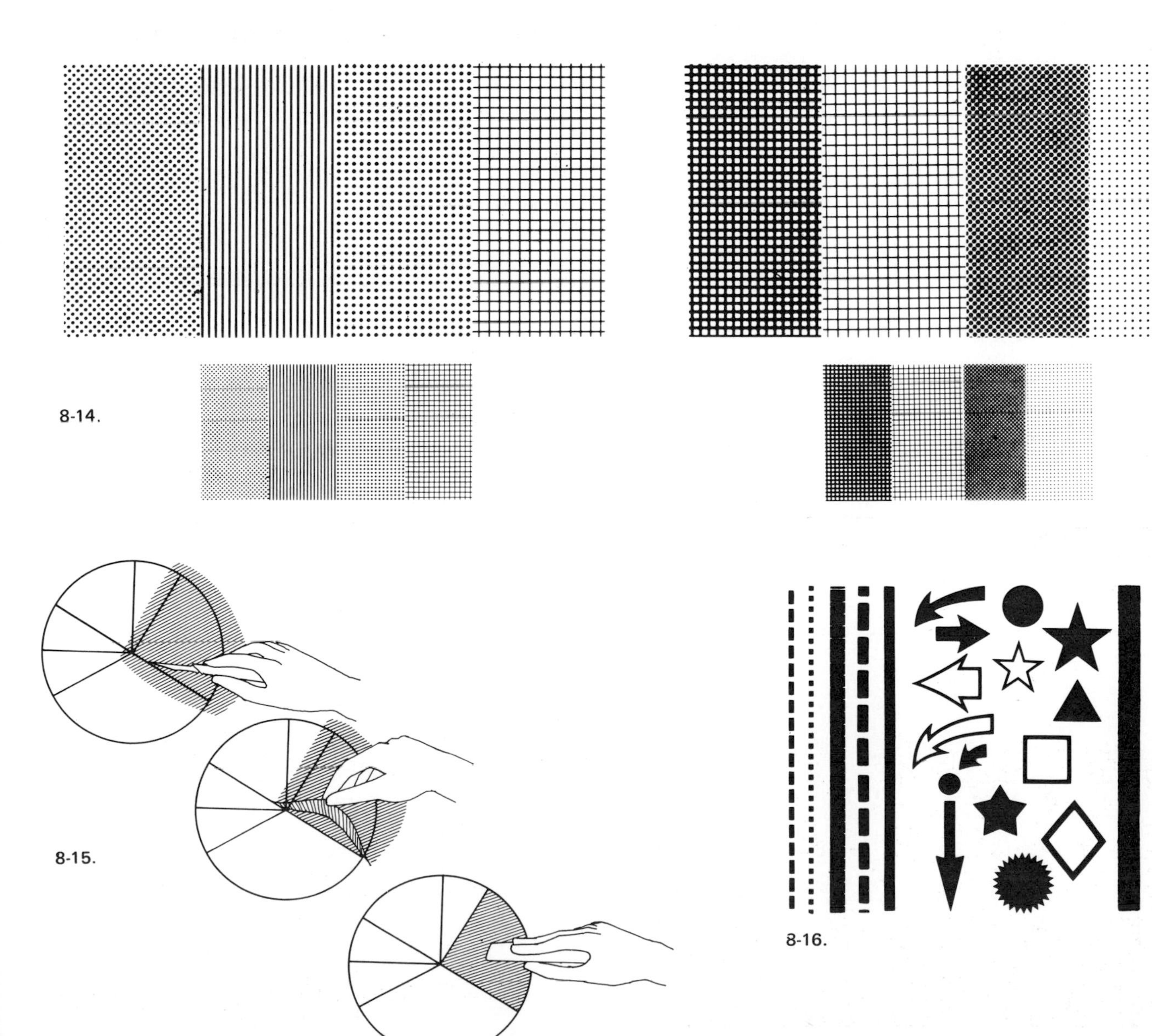

8-14.

8-15.

8-16.

Their crispness, variety, opacity, and consistent size make them invaluable for the creative designer of diagrams. Unlike ink, they are used up quickly, and nothing is worse than running out in the middle of a project, with none available at the store. Close inventory and a prejob check are necessary when depending on this method.

Tapes must be applied carefully to make sure that they are straight. When applying them, hold excess tape at either end for control. When the tape is in place, cut off this excess with a blade and burnish the rest so that it adheres to the ground. Curved lines can be formed with flexible crepe tape (8-17).

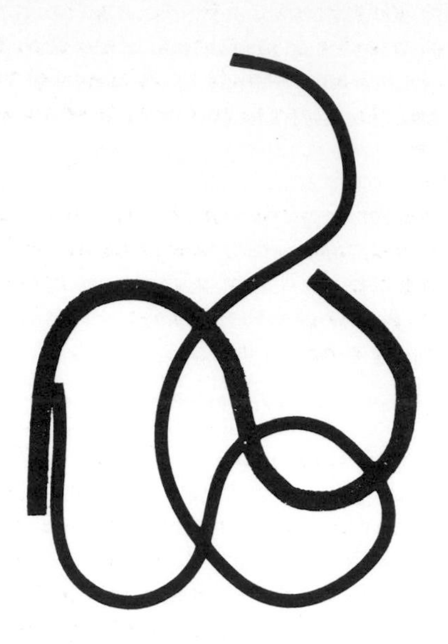

8-17.

Lettering

Lettering for diagrams is done in a simple sans-serif style. Titles or single words or phrases can be done in capital letters for particular emphasis, but lowercase lettering is read more quickly and is therefore preferable in most cases.

Labels are placed as close as possible to the subject that they name and should not extend outside the prescribed format proportion. Lettering should be planned with thought and deliberation as an element of the overall design. It should be arranged consistently: centered, flush left, flush right, under or over the subject.

Lettering techniques are discussed fully in chapter 9. Some quick recommendations for diagram lettering are shown here (8-18).

PHOTOTYPOGRAPHY: Optimum choice for looks, expensive, dependent on in-house machine or order from printer, black on white only.

DRY TRANSFER: Fairly fast, great style variation, crisp and attractive, expensive, dependent on close inventory check, black and color.

LETTERING GUIDES (LEROY): Fast, inexpensive after initial investment, few styles, many sizes, neat, black and color.

TYPEWRITER: INEXPENSIVE, FAST, ONE SIZE, OK FOR SMALL IN-HOUSE SEMINAR SLIDES, BLACK ONLY.

HANDLETTERING: FAST, HAS AN ATTRACTIVE IMMEDIACY BUT DEMANDS SOME SKILL, BLACK AND COLOR.

8-18.

PLANNING

The diagram is first carefully plotted or laid out in pencil. This is done on graph paper to ensure accuracy and to maintain the correct format proportion. It is then traced onto the final ground.

A drawing board is very important for straight alignment and for right-angled horizontals and verticals. A stable, heavy, metal T-square and a triangle are necessary to guide the pencil, pen, template, or tape accurately (8-19).

Ink system

Inking is a fast and flexible system that is not dependent on factors such as supply or availability. After the original investment the expense is limited to paper and ink. A wide size range of pens and lettering and symbol templates is available. With a little practice you can become comfortable and proficient with this equipment.

Using the Leroy scriber (pen holder) and pen point with a Leroy template, you can make horizontal or vertical lines or lines of lettering or symbols by sliding the template and scriber along the top edge of the securely held T-square (8-20). The adjustable scriber can also be adapted to form slanted lettering.

Most reservoir pens are designed to be used with the Leroy scriber and templates. Other lines are made with a standard pen holder and mechanical point held vertically to trace a beveled straightedge or curve.

Either an inked diagram or a photocopy can be sent to the publisher. If there is much detail, it is better to send the original.

Nonink system

Diagrams made with rub-on and dry-transfer letters, lines, and symbols are crisp and versatile, but they must be handled with care. The diagram should be mounted on a stiff backing to help prevent cracking or peeling off. They should not be exposed to a moist or cold atmosphere, as even slight moisture in the air may cause the design to curl away from the ground.

Color diagrams

Color diagrams for projection or display can be done directly by using color inks, tapes, and paper or indirectly by making a film positive and backing it with color board, paper, tapes, and paint. The film-positive method is more versatile, but the size of the transparency is not so flexible.

Direct method

When using translucent paper or film as the final ground, place the film over the pencil layout and trace the lines, symbols, and lettering directly. This can be done with Leroy pens and colored ink or adhesive-back tapes and symbols. Lettering can be done in either black or colored ink or dry-transfer letters. Any splicing, overpasting, or patching will show in the final slide or photograph and should be avoided. The same procedure can be followed when using one- or two-ply translucent paper on a light box.

When using opaque paper or illustration board as the background, the diagram is transferred onto the board using handmade or purchased transfer paper (Saral). The color elements are then applied over the penciled diagram.

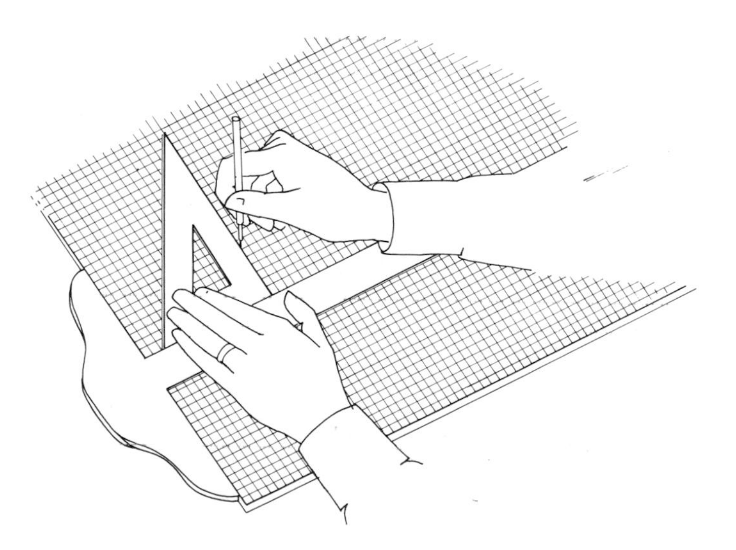

8-19.

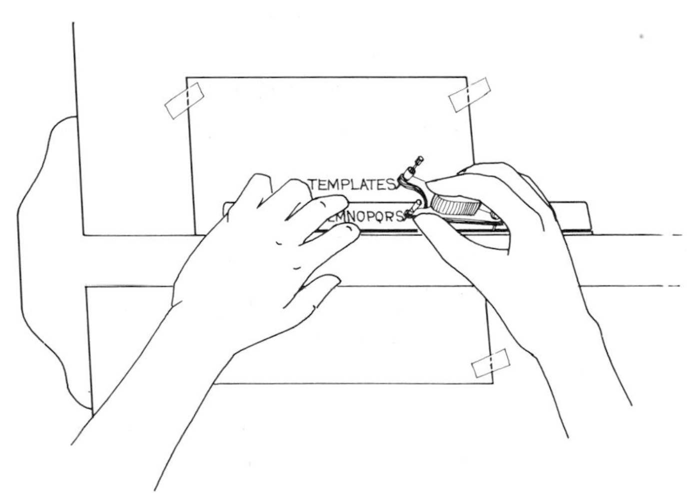

8-20.

Film-positive method

With the film-positive method the diagram is traced or pasted up in black on translucent or opaque white paper. It is then mechanically transferred to film, with only the black appearing on the clear film. Splicing and pasteup shadows do not show. This process can be done photographically, and the diagram either reduced or enlarged. The size is limited by your photographer's capabilities.

The transfer from paper to film can also be done very quickly in the studio with various copying machines. One system utilizes Transparex opaque film in the Thermo-Fax office copier and the Agfa-Gevaert Transparex washer. This method transfers your diagram at the same size (8-21). The film measures 8 1/2" X 11" (21.5 X 28 cm), and a margin of 1 1/2" (4 cm) is necessary on all sides for the photographic processing. If the original diagram is too big or too small, it can be enlarged or reduced on a photocopy or copy machine and the copy run through the machine.

The film positive is then mounted on a contrasting background of paper or illustration board. Color paper can be cut out and attached on the board under the film to highlight particular areas that you want the viewer to be most aware of (8-22). The color tells the audience where to look. Color tapes can be put on top of the film (crepe tapes for curved lines). Small areas can be backpainted with flexible vinyl-cel paint. Several coats are necessary to produce an opaque color.

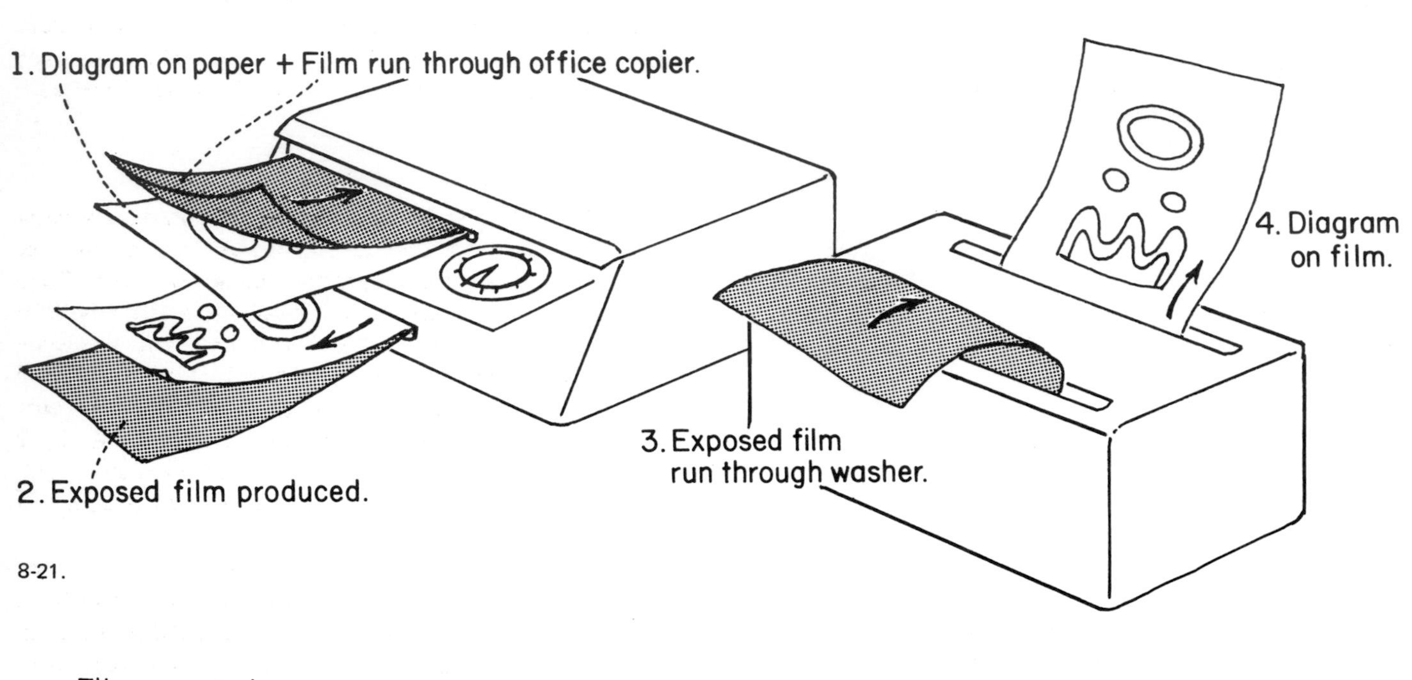

8-21.

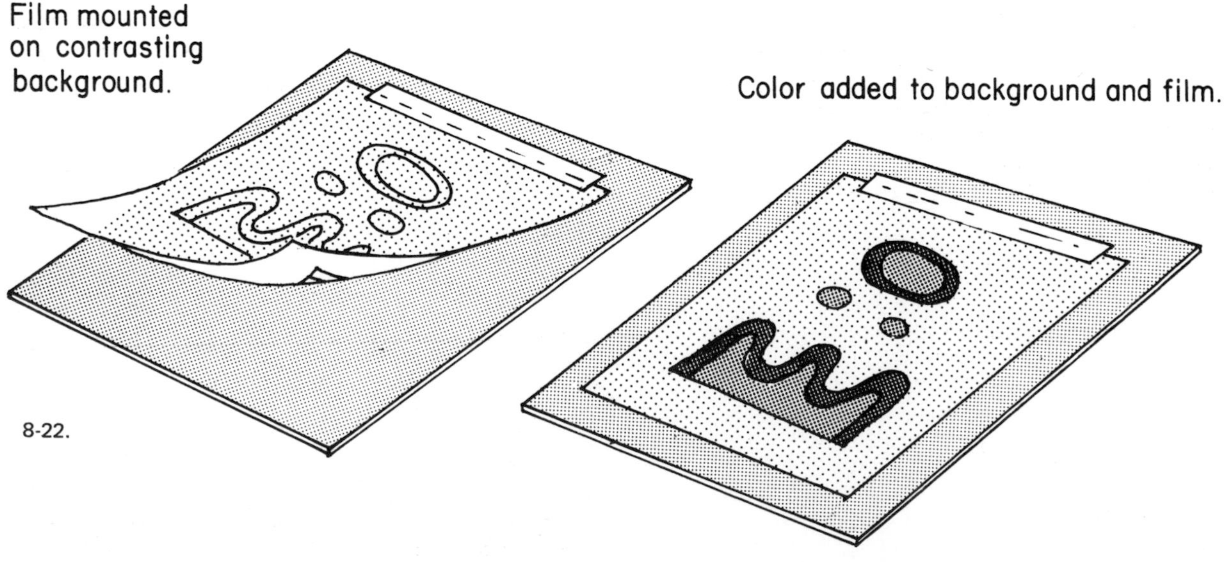

8-22.

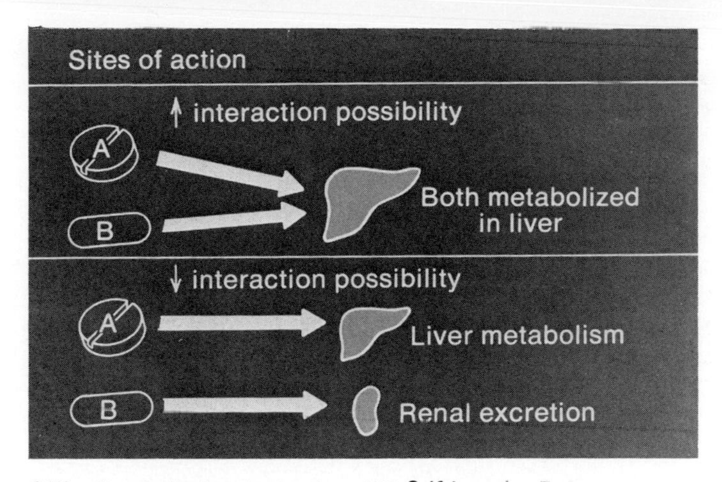

8-23. Factors in drug interaction. ACP Self-Learning Series.

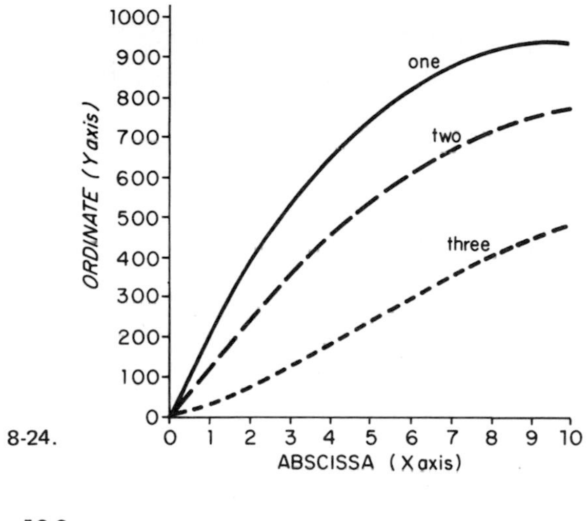

8-24.

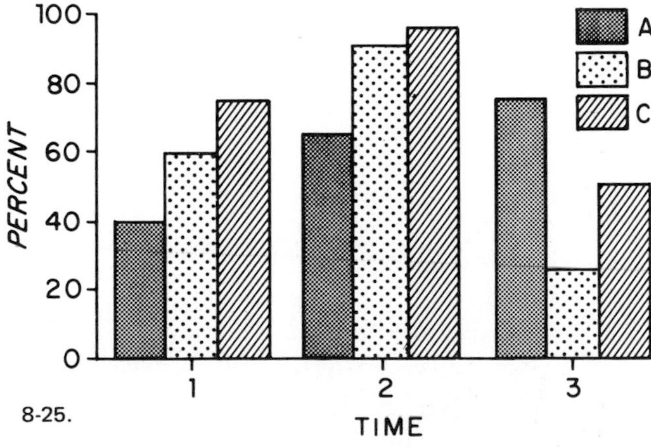

8-25.

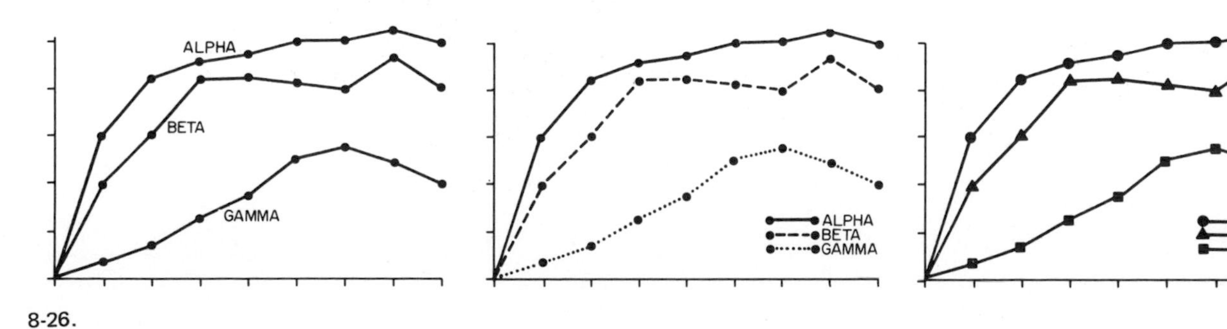

8-26.

The Transparex system offers the added versatility of color, as the film comes in ten colors including black and white. A white film positive mounted on a dark background looks especially crisp and carries well in a large auditorium. If several line colors are needed, they can be done separately on paper in black, run through the copier using different colors of film, and registered on a mounting board (8-23).

With the Transparex method all the copy must be the same thickness, so a pasteup must be photocopied before it goes through the machine. There must be carbon in the copy in order to reproduce, so felt-tip and ballpoint pens will not copy. Wax-back or dry-transfer letters will be destroyed by the heat of the machine if untreated. A thin dusting of talcum powder acts as a spacer between the dry-transfer lettering and the film to prevent the heat from destroying the lettering. An alternative is to use a photographic or machine copy.

A big advantage in using the film-positive method to produce color diagrams is that the original black-and-white version can also be used for publication, allowing double use of one piece of artwork.

Graphs

A graph or chart measures one value in relation to another value. The bottom horizontal-value scale is the abscissa, or x-axis. The vertical-value scale is the ordinate, or y-axis. These axes compose the constant part of the graph and should be thought of as the "frame" for the "picture." The lines, symbols, and columns that make up the "picture," or variable part of the graph, should be strong and clear. The "picture" lines should be several times thicker than the axis lines (8-24). Symbols should be large enough to be easily differentiated. Columns or bars should be filled with shading film or color for strength and solidity (8-25).

Quantitative parameters can be differentiated by labeling them or by using different lines or different symbols (8-26). It is not necessary to use both different lines and different symbols: in fact it may be confusing. Unless it is necessary to include the points for informative value, it is usually more effective to use variable lines and to omit the points. Scale lines inside the graph should be omitted unless they are necessary and should then be very thin (8-27).

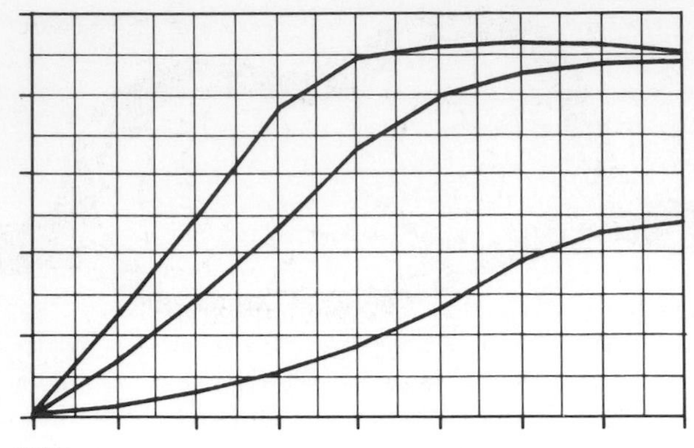

8-27.

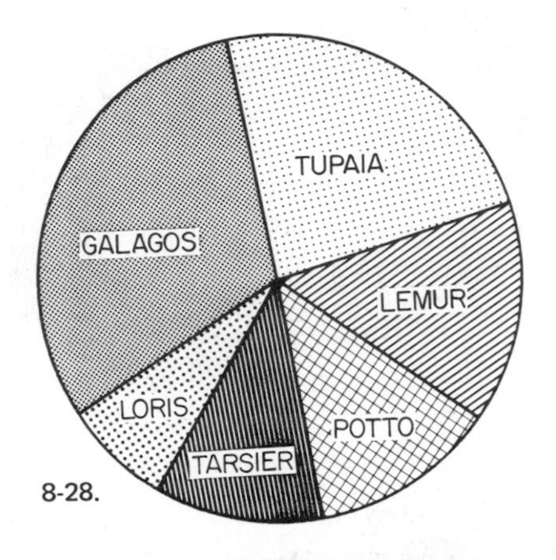

8-28.

The pie graph and some bar graphs compare quantities that are not measured against another scale. The pie graph should be limited to six wedges in variable shading film or colors. If possible, all labeling should be done directly on the appropriate wedge. It is particularly important to have good contrast in shading film or color between the areas (8-28).

Maps

As with diagrams, a map should be kept as simple as possible. Include only the physical characteristics of the territory that are necessary for the particular subject being studied and to orient the reader to the area. A road map would vary from a weather map in the details to be included. Both would be very different from a map showing the distribution of a species of animal or tree (8-29).

If states, cities, rivers, and mountains are to be labeled on the map, the lettering should show a progression of rank through the boldness and size of the labeling.

If necessary, an arrow pointing north and a scale in miles or kilometers may be included. While lettering is usually horizontal, it sometimes relates more easily if it follows the contour of the coastline, mountain range, or river. Shading film or color can be used to differentiate between land and water or other geographic variables (8-30).

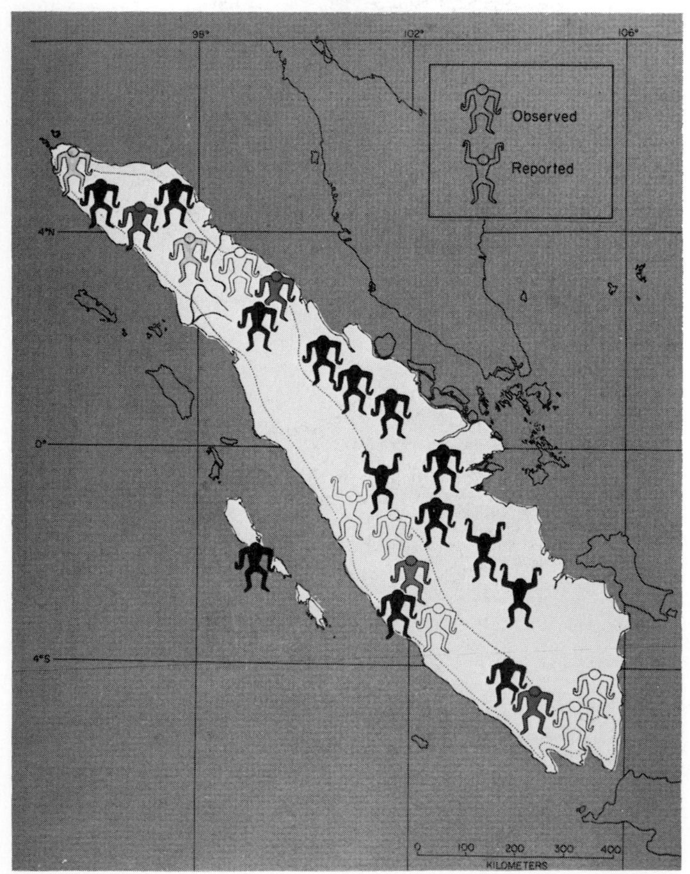

8-29. Phyllis Wood, designed for Carolyn Wilson. Gibbon distribution, Sumatra, Indonesia. Film positive, backpainted, crescent-board and cut-paper background.

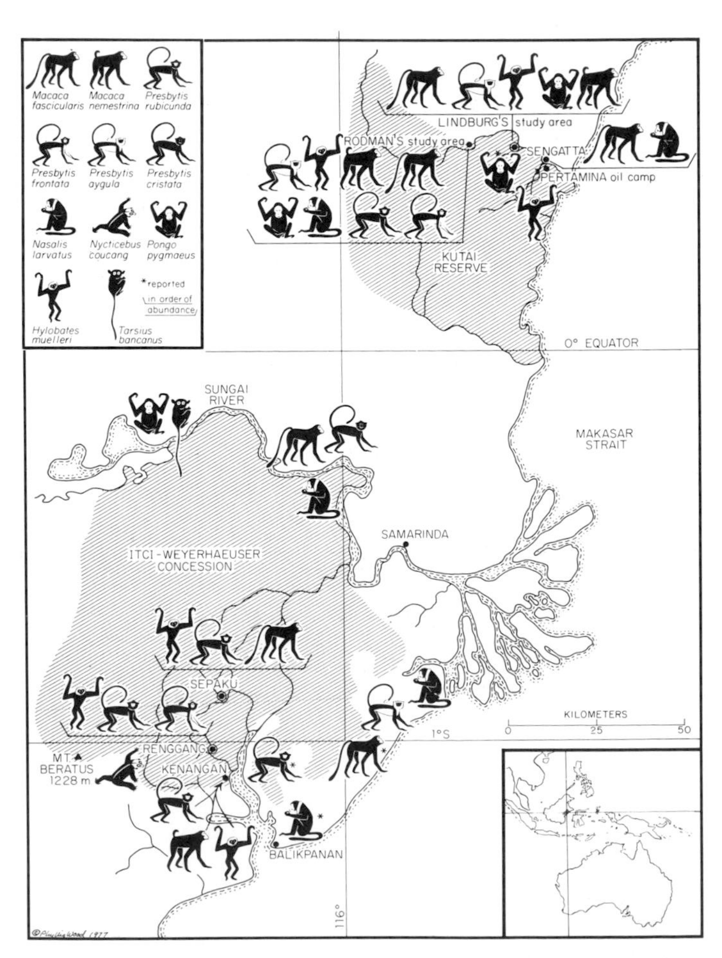

8-30. Phyllis Wood, designed for Carolyn Wilson. Monkey distribution, Kalimantan, Indonesia. Ink on paper with shading film and duplicated animals pasted up.

Schematic diagrams and abstract ideas

Schematic diagrams and abstract ideas should be reduced to the simplest terms (8-31). If several ideas are to be incorporated in one graphic, a gradual buildup from simple to complex is preferable to saturating the viewer with all the information in a one-step graphic. It is best to break the information up into separate steps, coordinated by a continuity of design and color (8-32). Design created by line direction, boldness, brightness, and contrast tells the viewer where to look and which is the important part of the diagram.

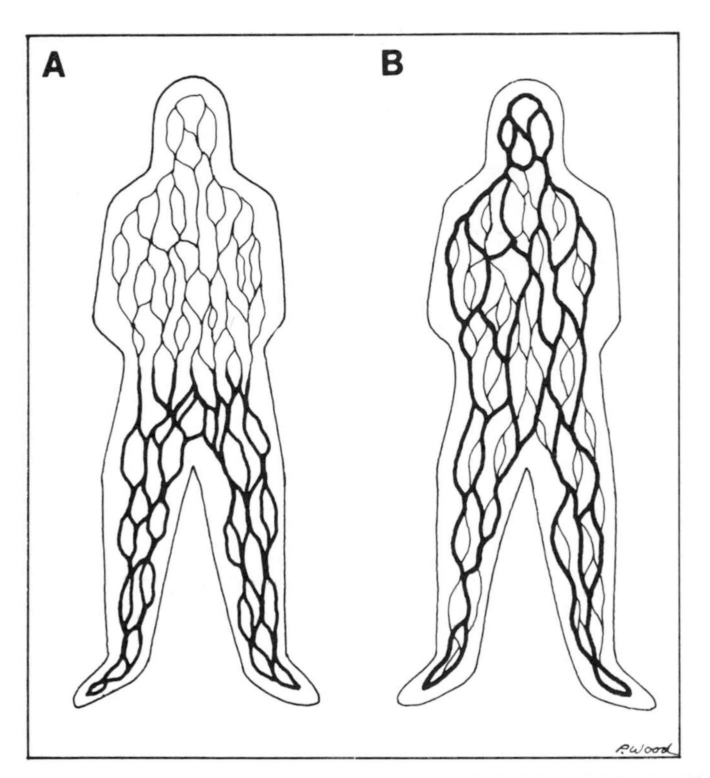

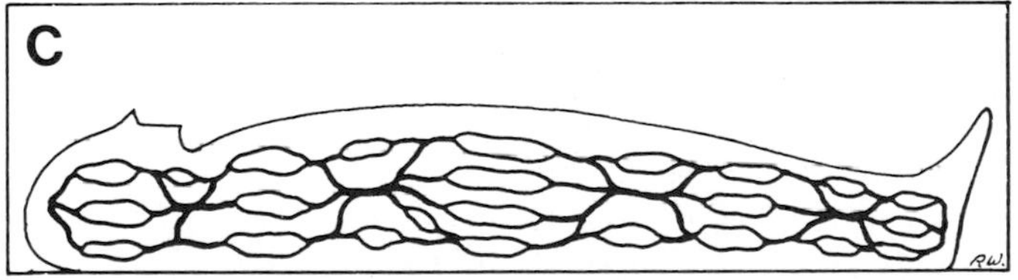

8-31. Phyllis Wood. From Cynthia J. Leitch and Richard V. Tinker, eds. *Primary Care*. F. A. Davis Co., 1978. Effect of gravity on vasculature. A. Uncompensated. B. Compensated by selective constriction. C. Passive compensation in supine position. Ink on paper.

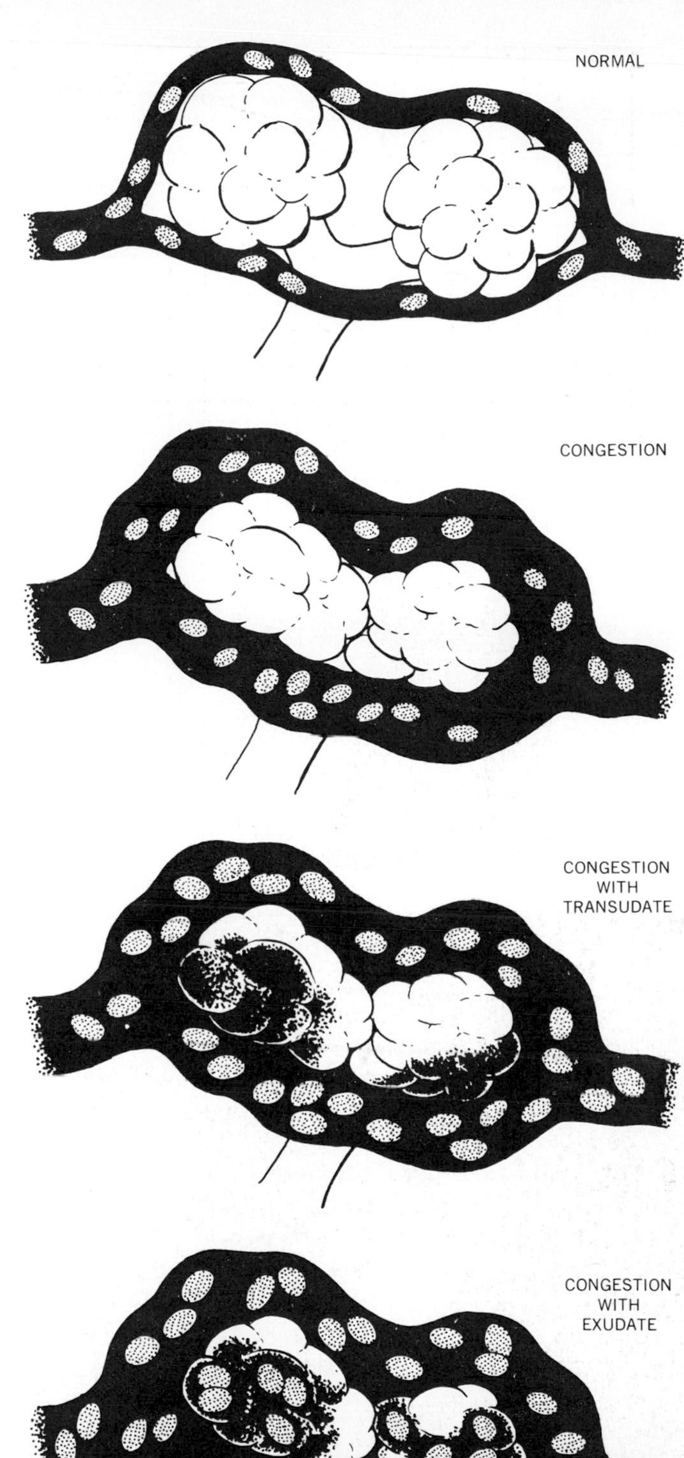

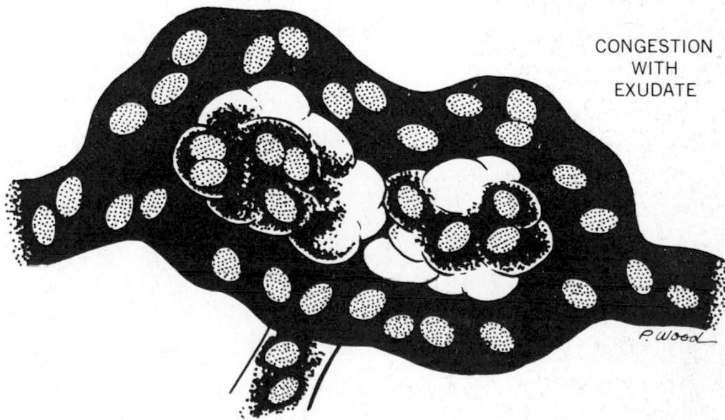

8-32. Phyllis Wood. Alveolar/capillary bed in lung, demonstrating egress of fluid from alveoli to capillaries. From Cynthia J. Leitch and Richard V. Tinker, eds., *Primary Care,* F.A. Davis Co., 1978.

CHAPTER 9.

Design and Layout

Design is the organization of graphic components into a unified whole. It includes drawings, photographs, labels and captions, borders and boxes, color, size, and proportion. Layout is the mechanical process of assembling, scaling, and pasting up these components in accordance with the printer's or photographer's specifications.

DESIGN

The *design* of a scientific illustration is part of the *rendering* of the illustration: the two elements cannot be separated from each other. While the scientist may look at an illustration and see only the subject matter, he sees it within the confines of the design, and the design influences his viewpoint at least subconsciously. Every illustration, no matter how simple, incorporates elements of design.

Scientific illustration is primarily concerned with communicating factual material, so the design should not be so overpowering, intricate, or detailed that it confuses the viewer or detracts from the communication value. Although the subject is factual, scientific, and serious, however, there is no reason why the design approach should be dull or static. The artist should strive to present the material in a creative manner while maintaining scientific accuracy and straightforward communication.

Every illustration should be designed with its ultimate use in mind. Journal publication, slide, and exhibit panel have different requirements as to size, amount of detail, and eye flow.

Source material

Go to the library and look through the newest graphic publications. The style and character of design change from year to year, and, although you may not be able to use any of the ideas directly, you will absorb the flavor of the new trends and be able to infuse a modern touch into your scientific design. Try not to stagnate in one style but rather to change and to grow with the times.

Recommended publications are *Graphis*, *Communication Arts*, *Graphic Design*, and *Novum.* Two annual publications are *Illustrators Annual* and *The Association of Illustrators Annual.* Graphis Press also publishes timely art books. Two that you will find helpful are *The Artist in the Service of Science* and *Diagrams.* Other publications such as *Scientific American* and *National Geographic* are good sources for innovative graphic interpretation of scientific material.

Elements

Three basic elements govern the organization and synthesis of design components: balance, eye flow, and size importance. All these elements should be considered when designing any graphic, whether it is large or small, a single illustration or a complex series.

Balance

Symmetrical or formal balance lends a feeling of stability (9-1), while asymetrical or informal balance has a more active or dynamic feeling (9-2). The grid is an orderly and methodical arrangement (9-3). Variations and combinations of these three basic arrangements depend on the desired emphasis and mood and on the natural display suggested by the subjects' shapes and the progressive sequence (9-4).

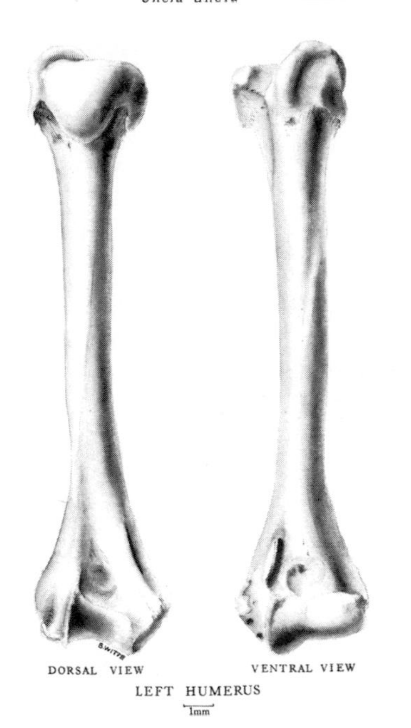

9-1. Beverly Witte. Wash on cold-press illustration board.

9-2. Beverly Witte. Wash on cold-press illustration board.

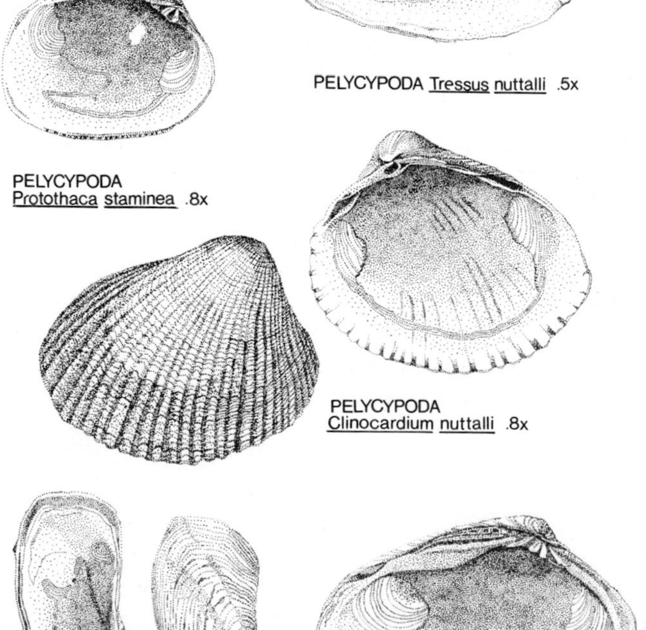

9-3. Maureen Huey. Ink on paper.

9-4. Lynn Bergelin. Ink on scratchboard.

Eye flow

The design should lead the eye through a natural progression of the information. The strong lines or thrusts of the drawings and the arrangement of the other elements guide this direction.

The most important part of the printed page is the upper left, the area at which the viewer first looks. Take advantage of that fact when designing for print media (9-5). In projected media the entering point drifts down to center left. After the viewer's eye has been attracted to the spot, the design should guide it in the correct eye-flow direction. The natural progression of the illustration should be indicated by strong lines or thrusts in the design (9-6). If a spiraling direction of eye flow is the correct viewing sequence, the dominant design direction should indicate that order (9-7). No strong lines or thrusts should lead away from the design or off the page or be left dangling.

Size-importance scale

Each part of the illustration should be assigned a value on the size-importance scale. This is usually determined by the amount of detail that must be included and not by the actual size of the subject. The gestural pose of an animal shown with little detail may be drawn small, while the face, if shown in much detail, may be high on the size-importance scale (9-8). If subjects are equal on the size-importance scale, they should be treated equally (9-9).

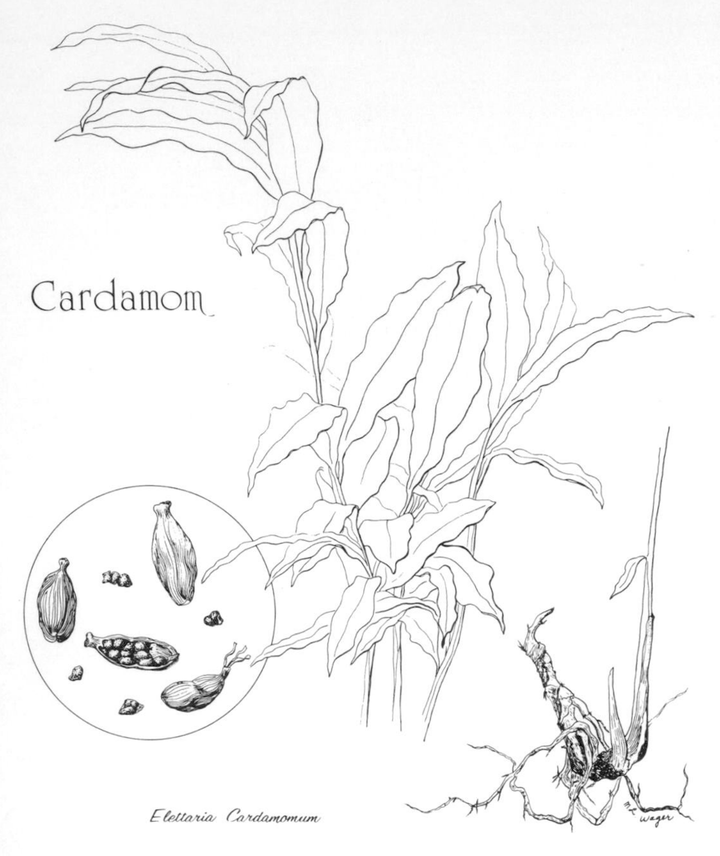

9-5. Mary Lou Wager. Ink on paper.

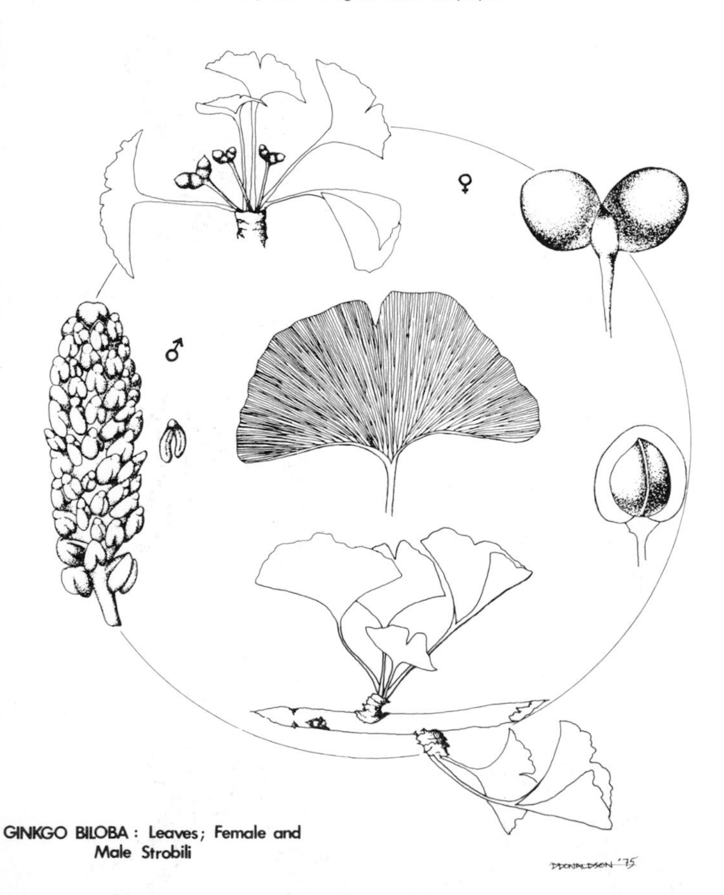

9-6. Deborah Donaldson. Ink on paper.

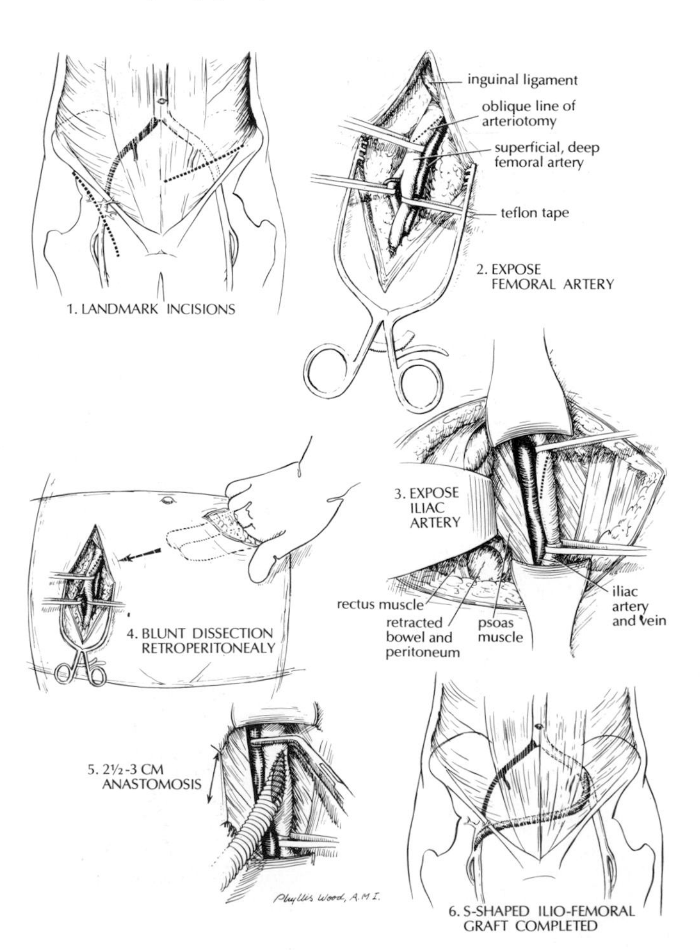

9-7. Phyllis Wood. Ink on scratchboard.

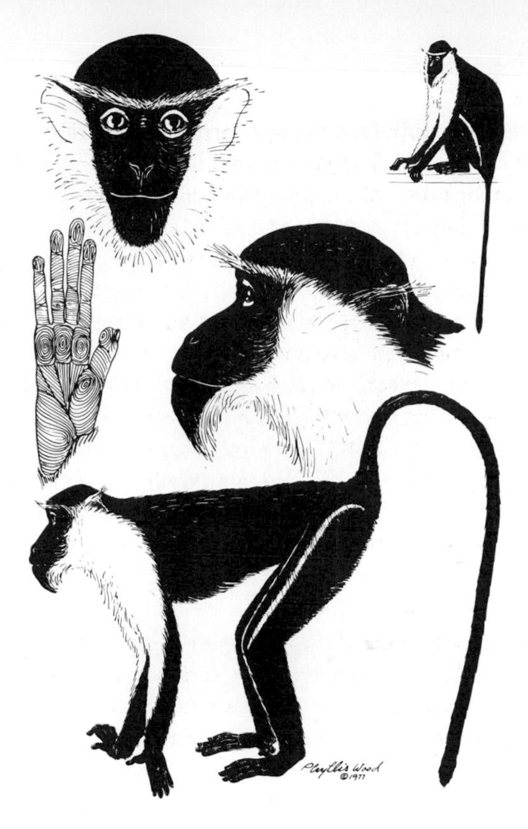

9-8. Phyllis Wood. ***Cercopithecus diana*** (Diana monkey). Ink on scratchboard.

9-9. Phyllis Wood. Monkeys. Ink on scratchboard.

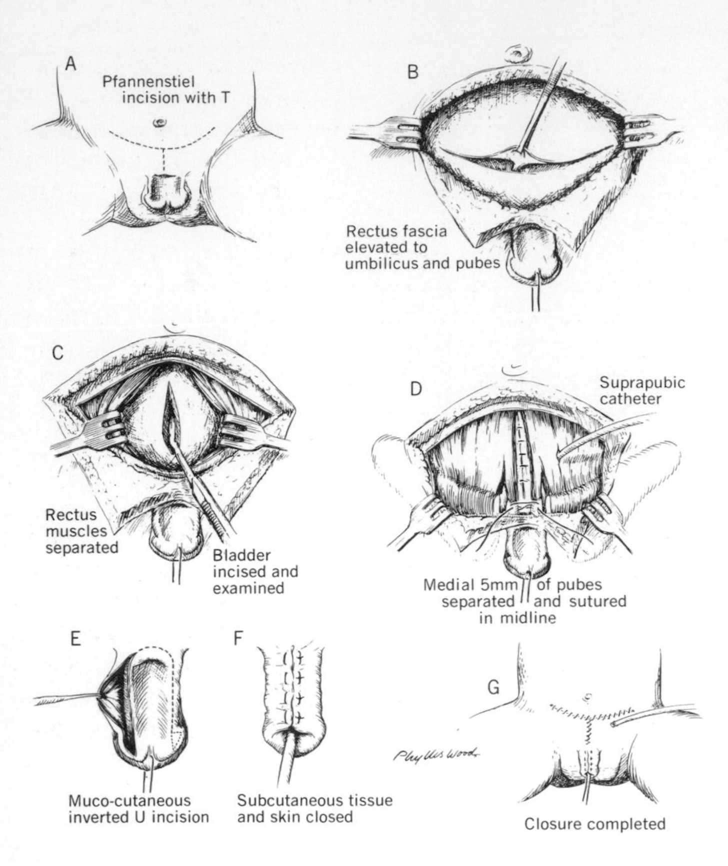

9-10. Reconstruction of urethra in infant. From Julian Ansell. *Urologic Surgery.* Harper and Row.

Lettering

Lettering is considered as part of the design and should be planned for style, size, and placement as carefully as is the rest of the illustration. The placement of the lettering can be an important design factor in eye flow.

Arrangement

Labels and captions should be placed close to the elements that they describe so that the leaders are as short as possible or omitted completely (9-10). Arrange the labels in a compact manner, as the greater the outside dimensions are, the more reduction will be necessary to fit the illustration into its assigned space (9-11).

Leaders (or call outs) are thin solid, dashed, or dotted lines. They should not cross each other. They can be black, white, or a combination so that they contrast with the illustration.

Labels should be arranged in a meaningful way. They can be all flush left (even left-vertical margin) and/or flush right (9-12). Flush left is easier to read, which is especially important if there are many labels. Leaders can all start at the same distance from the drawing (9-13).

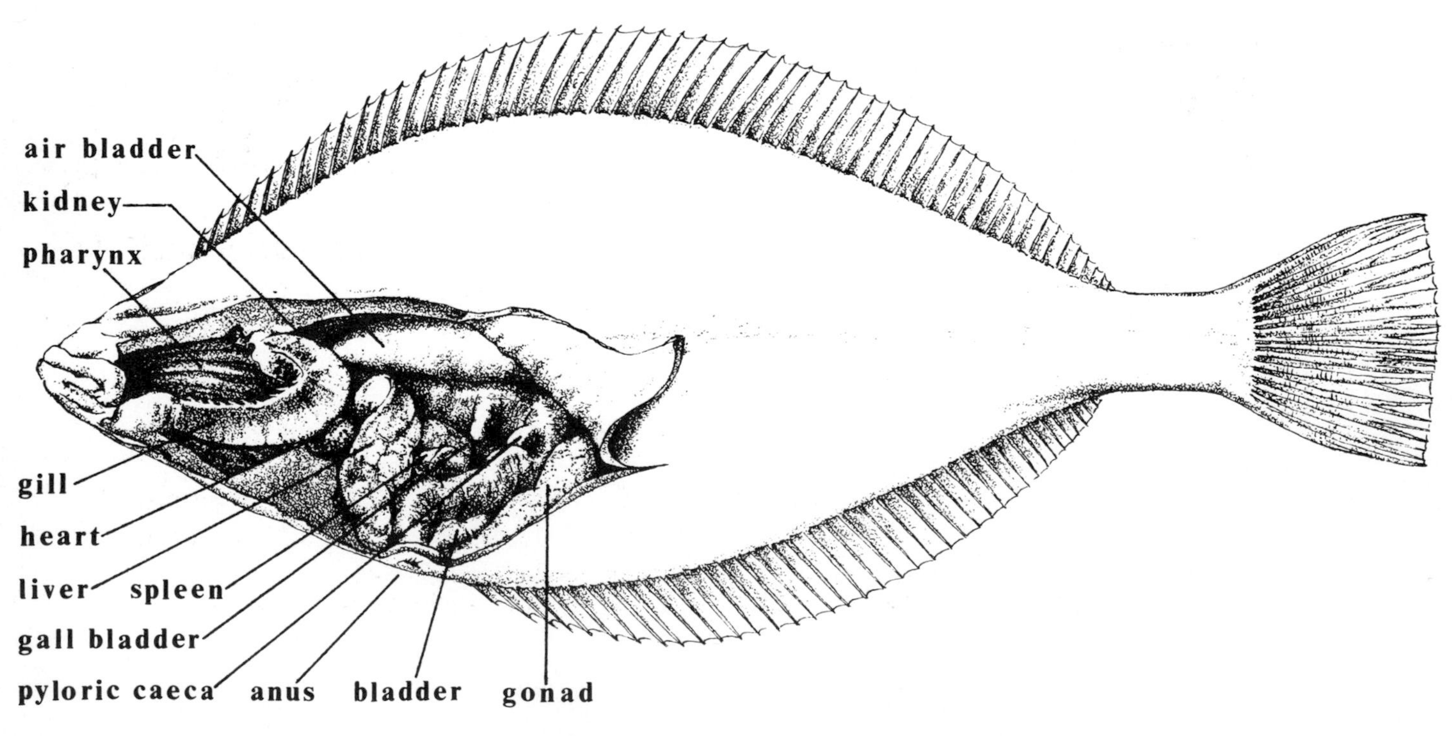

9-11. Anton Friis. *Hippoglossoides elassodon* (flathead sole). Litho crayon on coquille board.

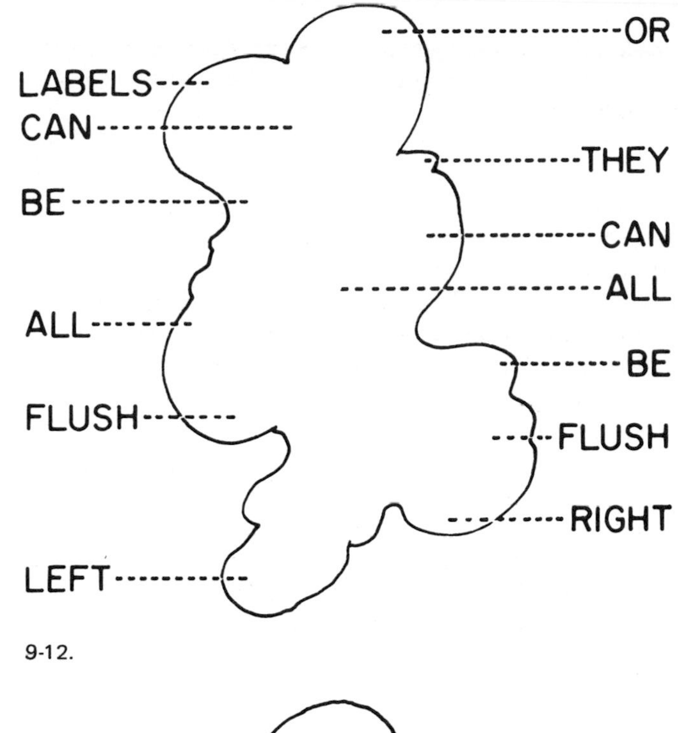

9-12.

If there are too many words in a label unit to fit on a single line, the unit should be broken into several lines, taking into account both design and ease of reading. Phrases should be read aloud to determine the best place for the pause that indicates a new line (9-14). Hyphenated words are rarely necessary and should be avoided.

Abbreviations are to be avoided unless they are familiar to and commonly used by the reader or unless space is so tight as to require them. In the latter case they should be explained in the figure legend. Letters used as labels and keyed to the label names should be avoided. Illustrations should be as complete as possible, with all pertinent information included in the most easily understandable position.

Style

Sans-serif lettering is usually preferred in scientific graphic design, as it is straightforward in mood, uncompromising, and unaffected (9-15). Its personality does not detract from, alter, or add to the facts presented. It does not have thin serifs that may disappear in photography or projection. Body copy with serifs is considered faster to read, but the small amount of lettering on an illustration is not affected by this time factor.

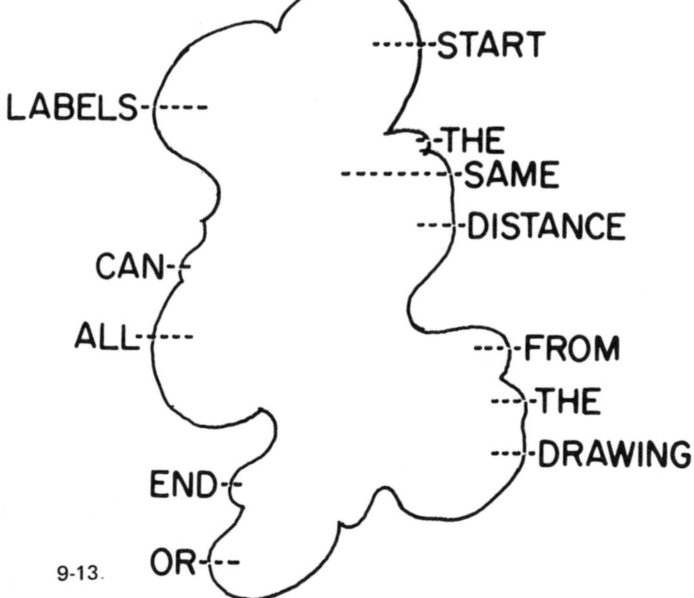

9-13.

Helvetica is a sans- (without) serif face.

Times Roman is an example of a serif face.

9-15.

Phrases should be
read aloud to deter-
mine the best place
for the pause that
indicates a new line.

Phrases should be read
aloud to determine the
best place for the pause
that indicates a new
line.

Phrases should be read aloud
to determine the best place
for the pause
that indicates a new line.

9-14.

Size

Lettering size is determined by the size-importance scale within a range specified by its ultimate use.

Lettering for publication should be approximately the same size as the body copy. If it is much larger, it will have an unpleasant, gross appearance. If it is much smaller, it will be annoying to the reader and hard to decipher. Lettering in projected illustrations should be relatively larger so that it can be read easily. For television lettering should be at least 1/14 the height of the screen in order to be easily legible (9-16).

Lettering size is measured in points; lines of type are measured in picas (9-17). There are 12 points in 1 pica and 6 picas in 1″ (2.3 picas in 1 cm). Typefaces are measured in points from the top of the ascenders to the bottom of the descenders (9-18).

Upper- and lowercase

All capital, or uppercase, lettering can emphasize words or titles, making them higher on the size-importance scale, but lowercase lettering is more readable. It is particularly important to use lowercase lettering for extensive labeling or captions. There is a descending hierarchy of caps and lowercase: (1) all caps, (2) initial caps on all words, (3) initial caps on important words, (4) initial cap on first word, (5) all lowercase (9-19). This sequential importance scale shows the viewer what to read first.

Bold or thick lettering also affects the size-importance scale. Thick or bold lettering should be examined critically before it is used. Thick lowercase a's, o's, and similar letters may close up when reduced. The body copy in the text indicates the lettering thickness that is comfortable for easy legibility (9-20).

Optically even spacing

Letters should be spaced so that they are optically equidistant from each other. Consider *all* the space between the letters (9-21). If you poured sand into the spaces, they should all be equal. Following this rule, o's and other round letters would almost touch each other, while vertical letters would be further apart. Each word should stand as a unit, not as a string of letters. The spaces between the words should merely separate the words: they should not break up the horizontal sequence of the phrase. In order to produce optically even lettering—letters that appear to be lined up evenly although they are actually uneven—some letters must be adjusted slightly. Letters that are round at the bottom, such as O, C, and S, should be placed slightly lower than those with horizontal or vertical bases, or they will appear to bounce up (9-22). With a flush-left margin letters such as O, C, V, and T should be moved slightly to the left of the margin of vertical-left letters (9-23).

TELEVISION LETTERING
SHOULD BE
AT LEAST
ONE-FOURTEENTH
SCREEN HEIGHT

9-16.

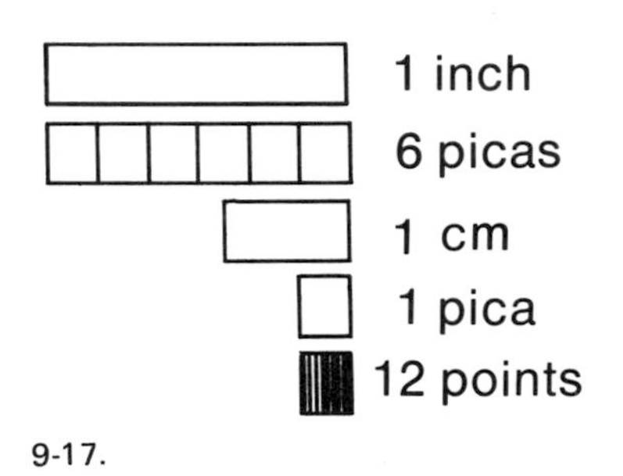

9-17.

top of the ascender
Typeface
48 points
bottom of the descender

9-18.

ALL CAPS
Initial Caps Of All Words
Initial Caps of the Important Words
Initial cap of the first word
all lower case

9-19.

LOOSE

9-22.

HELVETICA LIGHT

HELVETICA MEDIUM

HELVETICA BOLD

9-20.

DON'T MOVE
VERTICAL
LETTERS
OVER
TO THE LEFT

9-23.

Coolly Coolly

9-21.

Lettering methods

Methods of lettering vary in appearance, cost, and convenience. Selecting the cost-effective method for each job is part of the illustrator's responsibility.

Phototypography

Phototypography is the most consistent, crisp, and clear kind of type. It is produced by a photographic process from different-style fonts. A font is similar to a typewriter keyboard in that it includes an upper- and a lowercase alphabet, with numbers and punctuation marks in a specific typeface and point size. It may also include italic, bold, fine, extended, or condensed styles (9-24). The lettering is cut and pasted directly on black-and-white line illustrations and on overlays for continuous-tone or color illustrations (see chapter 10).

The machines that produce phototype vary from complex computerized operations located in a printing plant to in-studio machines that are operated by keyboard or disk. Lettering is available in a large range of sizes and typefaces. Spacing between the lines of type and between the words and letters is adjustable. Studio phototypositors usually produce type in strips containing a single line of type. The print-plant typographer returns your type to you on sheets of paper, arranged according to your specifications.

When ordering type from a printing plant, specify your type requirements carefully so that the typographers do part of your work for you. The type comes back to you ready to paste up without time-consuming adjustments. Type the copy and indicate the point size and typeface. Check it carefully, as the typesetters do not correct misspelled words or incorrect punctuation and capitalization. Note words that are to be set in boldface or italics. Indicate which units of type should be set flush left, flush right, justified, or centered (9-25).

Spacing between the lines should be noted, or the type will be set solid. Type that is set solid with no extra leading (space added between the lines) is usually uncomfortably close together. Leading (pronounced "ledding") is measured from the base of one line to the base of the next line. In solid type the leading is indicated by the same number as that of the type's point size: 12-point type set solid would be indicated as "12 on 12." Solid type leaves only a hairline space between ascenders and descenders. Adding 1 or 2 points of leading (12 on 13 or 12 on 14) between the lines is usually sufficient for legibility (9-26). A type gauge (Haberule) is invaluable in determining the leading desirable for a particular job.

Dry-transfer lettering

Dry-transfer or rub-on lettering is available in an almost unlimited variety of typefaces and is therefore used for many specialty jobs. It is fairly expensive and dependent on an efficient inventory system. It is comparatively slow to apply. Its main appeal lies in its crisp good looks and its range of styles, and there is no waiting for the type order (9-27). In phototypography margins and base lines are done for you, but you must do these steps with dry transfer letters. It is important that the letters are optically level and evenly spaced.

Italics, **bold,** light, extended; condensed

9-24.

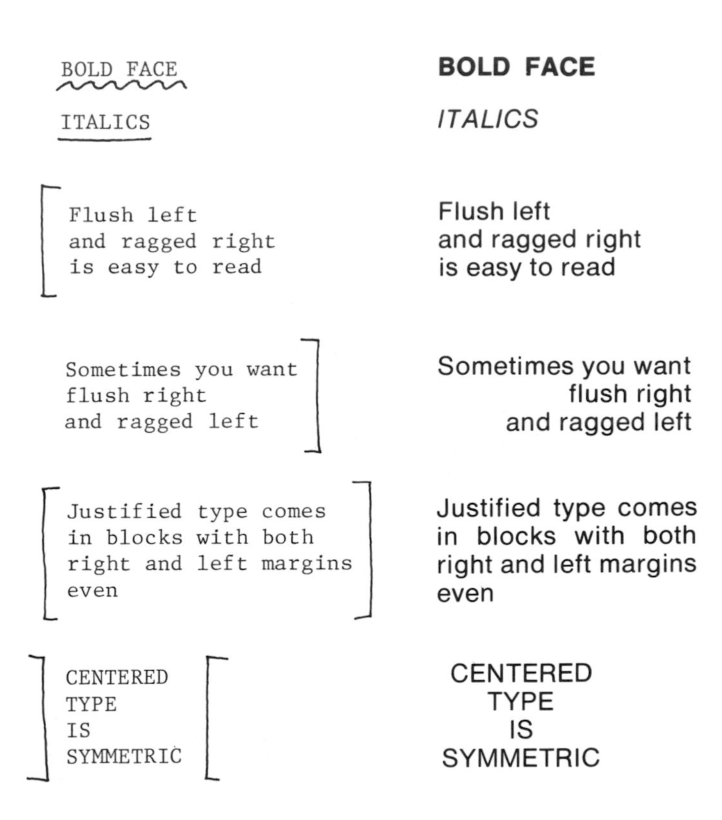

9-25.

Type that is set 12 on 12
is set solid with no leading
between the lines.

Type that is set 12 on 13
has one point of leading
between the lines.

Type that is set 12 on 14
has two points of leading
between the lines.

9-26.

Dry-transfer lettering comes in black, white, and some colors, so it can be used on both light and dark contrasting backgrounds. It can be used directly on color or continuous tone or on an overlay. It can be used to label photographs, contrasting with either the light or the dark values.

Apply it on a hard, smooth surface, rubbing the letter with a burnisher or other hard, smooth object until the entire letter is gray, which indicates that it is transferred. Take care not to crack or split the letters. Burnish them down thoroughly through a protective piece of paper. Spraying them with a mat fixative helps to preserve them, although they are not dependably permanent.

Leroy lettering system

The Leroy mechanical system of producing inked lettering is fast and inexpensive after the initial purchase of the lettering set. The lettering can be done directly on black-and-white line copy, on continuous-tone drawings or photographs, and on color backgrounds. It has a neat appearance and offers a broad range of letters and line thicknesses, but it is not as crisp and professional-looking as phototype or dry-transfer lettering (9-28).

Handlettering

Handlettering is usually not considered to be consistent with a professional presentation, though it has certain casual charm in quick slide presentations. Only an expert calligrapher should attempt handlettering on an illustration meant for publication.

Typewriter

Most office typewriters produce type that is somewhat gray, and each letter, number, and punctuation mark is given equal space. This letterfit arrangement is not satisfactory for a professional look. Some typewriter compositor machines such as the IBM Selectric Composer or the VariTyper produce black type and assign space to characters according to their size. The largest type size, however, is 12-point, which is not big enough for an illustration that will be reduced.

LEROY TEMPLATE SIZE 140 • PEN SIZE 0

Template Size 200 • Pen Size 1

Template 290 • Pen 2

9-28.

Avant Garde GOTHIC MEDIUM

Helvetica Medium

Univers 53

Bulletin TYPEWRITER

Friz Quadrata

Bookman Bold

Optex

Pump

9-27.

Drawing size

Drawings are usually rendered one and a half times larger than they are to be printed. A drawing with a great deal of detail may be drawn twice as big as the printed illustration (two up). Drawing the original larger than printed size allows the artist to include details that would be very difficult to render at the printed size. Reduction improves a drawing's quality, diminishing slight imperfections and intensifying crispness. It is preferable to be consistent in the reduction requirements throughout a single publication so that the printed quality will be consistent. In any case an illustration should never be enlarged, as this emphasizes any imperfections and turns a good line into a ragged one. A drawing should be planned for a specific number of inches or picas, for one or two columns, or for a certain percentage of a page (9-29). Instructions to the printer for sizing your illustration should be written under the drawing, with a bracketed line marking the area to be reduced. It can be labeled "reduce to 3″," "reduce to full-page width," "reduce to one column," or any other desired reduction. If the page size is not known, it may be preferable to write the reduction instructions in percentages. This should be labeled as "reduce to 60%." Do not write "reduce 60%," as this may be interpreted in several different ways.

Scaling

In scaling there are two elements to consider: the shape and size of the illustrations and lettering and the proportions of the page, page portion, slide, or TV screen that you are using.

Diagonal lines

Assume that you have a full page in a journal for your presentation. Measure the size of the printed page. Consider only the part of the page that has printing on it, excluding the page number (folio), the chapter title, and the margins. This is the area available for your illustrations. Draw this rectangle on a large blank sheet of paper and then draw a diagonal line upward from left to right, bisecting the opposite corners. Extend the left-vertical border and the bottom-horizontal border. Using any point on the oblique line as the upper-right-hand corner, you can create a rectangle of any size that will have the same proportions as the original page size (9-30).

Proportional scale

The proportional scale is a handy, inexpensive tool made of cardboard or plastic that gives you the unknown fourth measurement and the percent of reduction or enlargement.

Suppose that your printed-page size is 6″ X 9″ (15 cm X

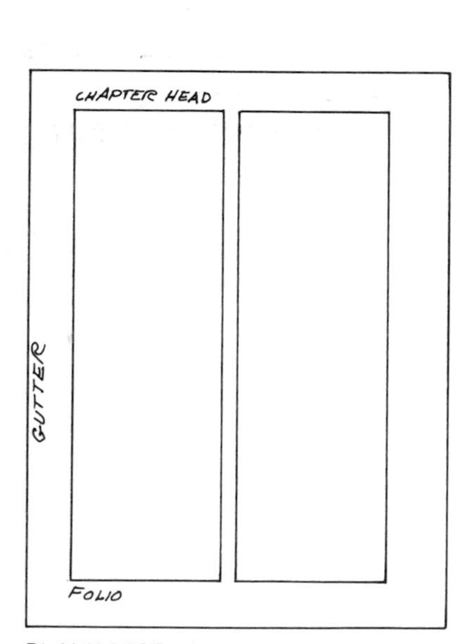

BLANK PAGE LAYOUT

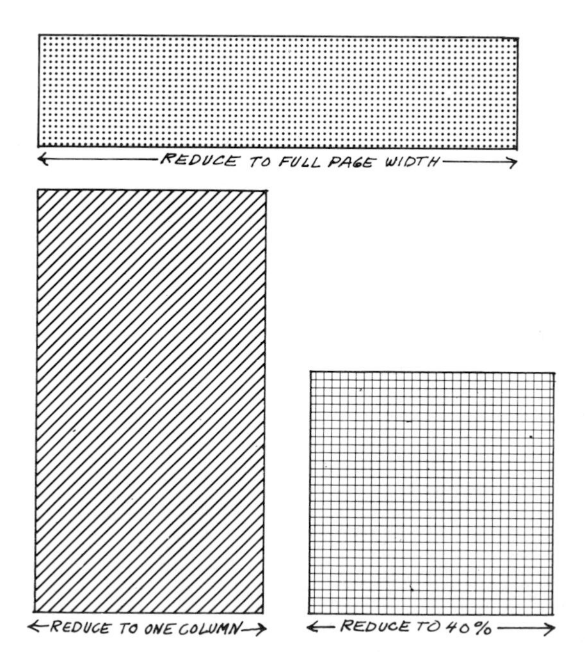

SCALED ILLUSTRATIONS

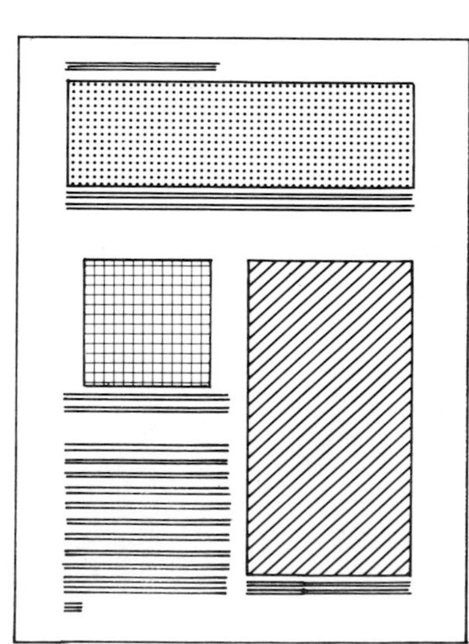
FINAL PRINTED PAGE

9-29.

23 cm) and that the longest horizontal element in your illustration is 11″ (28 cm). You want to know the vertical measurement that will take full advantage of your page proportion. Using the proportional scale, match up 6″ (15 cm) on the reproduction-size scale with 11″ (28 cm) on the original-size scale. On the reproduction-size scale 9″ (23 cm) matches up with 16 1/2″ (42 cm) on the original-size scale. This indicates that the optimum vertical measurement is 16 1/2″ (42 cm). At the same time it indicates that the drawing will be reduced to 54%. This is an accurate indication for the size of your drawings, the weight of the lines, and the type sizes (9-31).

Arranging the elements

After planning your type and illustrations and determining the area size that you are committed to you should coordinate these elements with the balance and eye-flow direction that you want. Mark off a rectangle of the appropriate size. Make machine copies of the penciled illustrations, the scaled photographs, and the type and cut them up. Arrange them in the most natural and meaningful way on the area that you have marked off. Try several different arrangements. Some of the elements should perhaps be reduced or enlarged for more or less emphasis or so that they will fit in the format (9-32).

The cut edges of the various elements make them appear to be closer together than they in fact are. Put a sheet of tracing paper over the entire layout and squint at it to filter out the cut edges from your vision. This gives you a clearer idea of the actual appearance of the drawing in print or on a slide projection.

If color is to be used, it should be rendered directly on the layout in color pencil or felt-tip marking pen, using the actual planned colors.

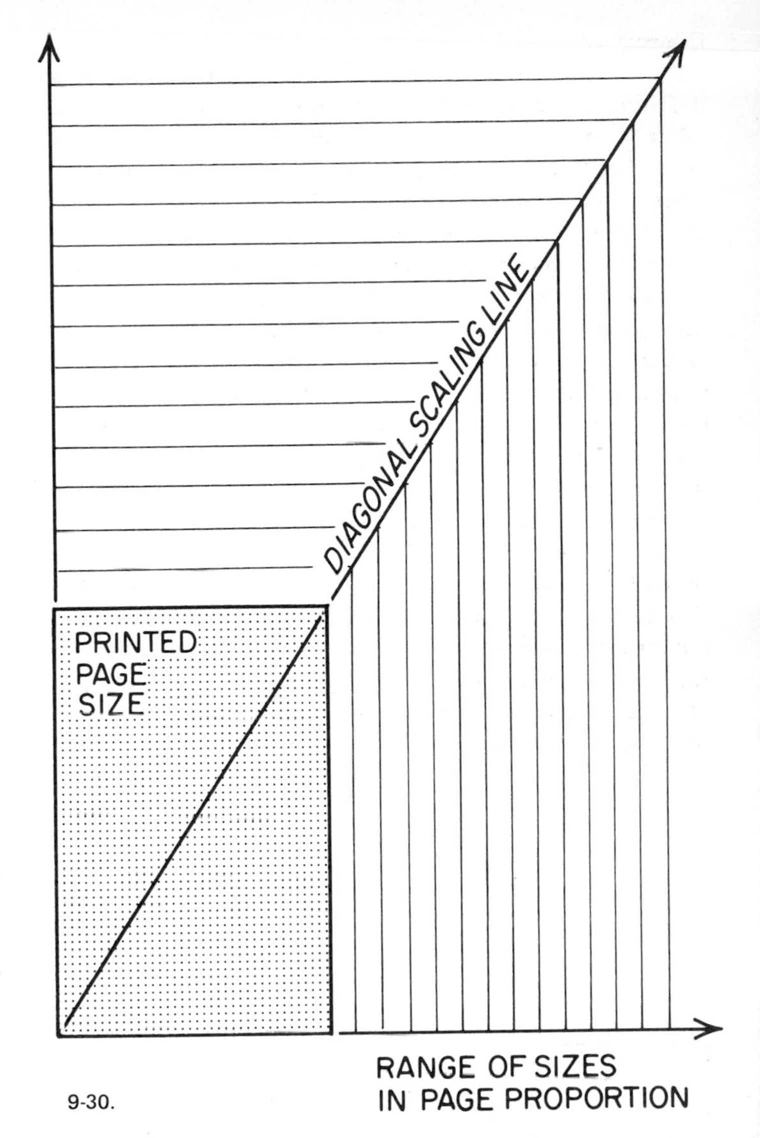

9-30.

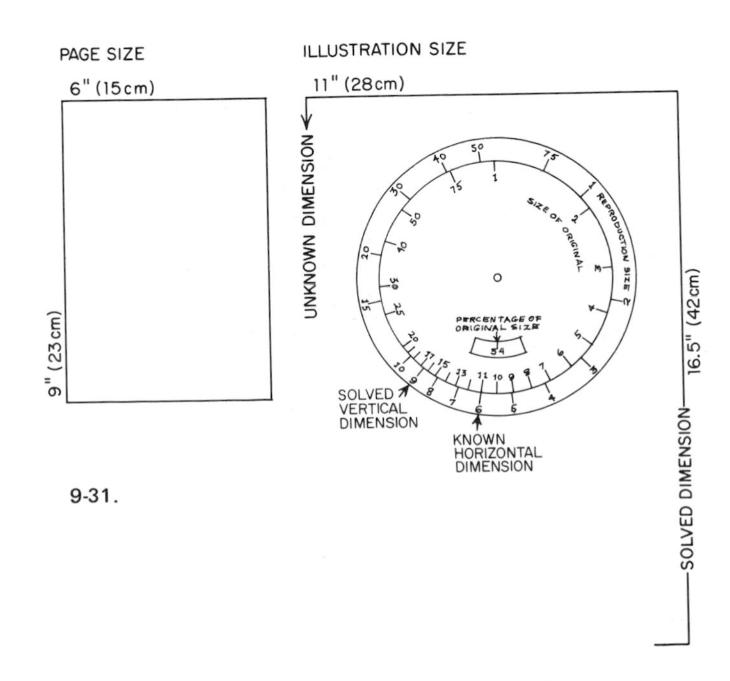

9-31.

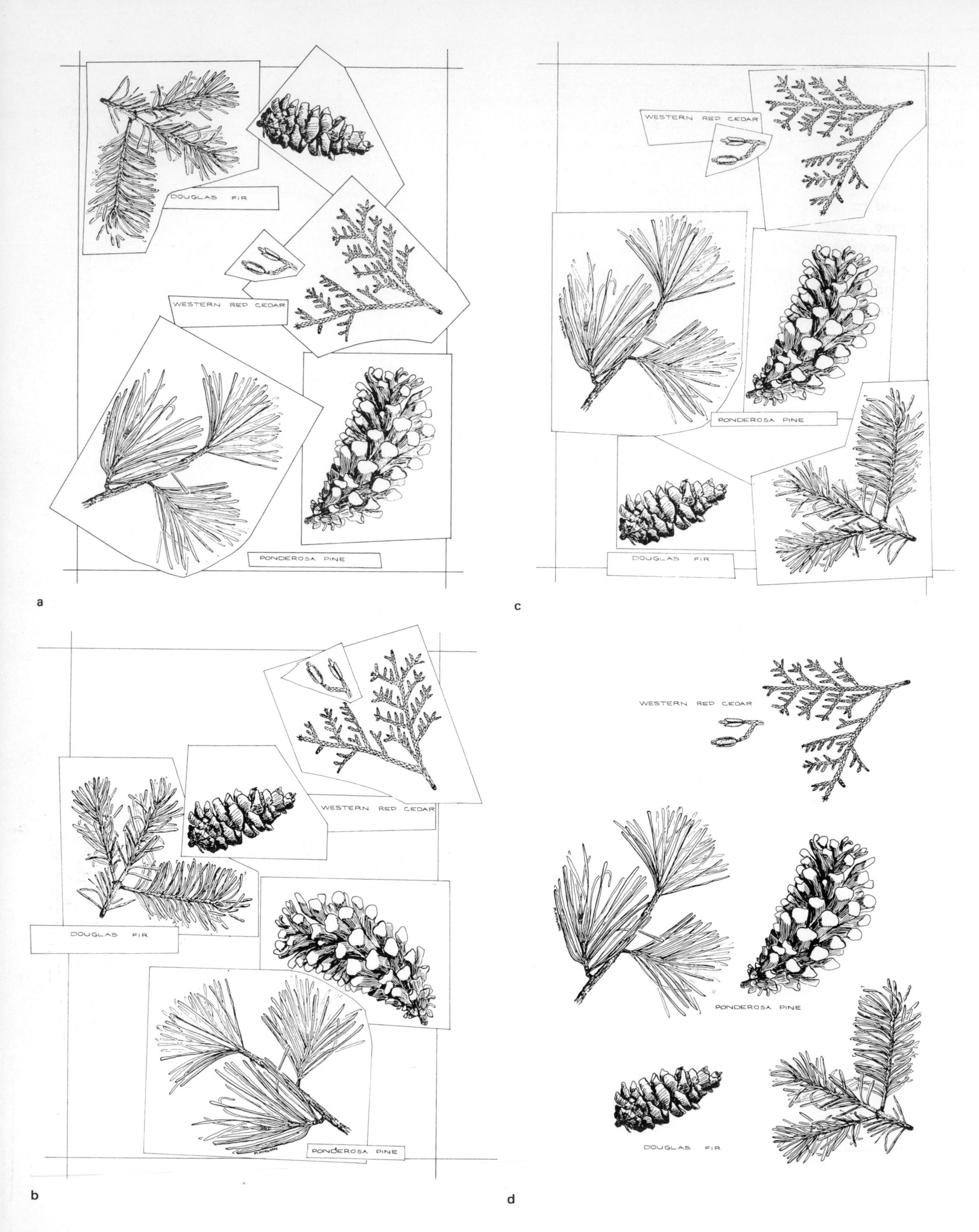

9-32. Molly McMurray Coulter. Ink on paper.

Borders

Borders used in scientific illustrations are usually simple single or multiple lines. They are used to unify elements in the illustration, to guide the eye, and to keep the various elements from "falling off the page." A border can be a single rectangular box around the drawing. It can enclose part of but not all the drawing. It can have rounded corners or be circular. It can be used to spotlight only one part of the illustration. It can be used to separate the drawing from the text or from other drawings (9-33). The border can be a very effective and simple design element. Put a sheet of tracing paper over your drawing and try different borders to see which most enhances your illustration.

Final layout

To assess your design, ask yourself whether, if you were unfamiliar with the subject illustrated, it would now be clear; whether the main emphasis of the message is easy to recognize; whether the sequential order is natural to follow; whether the design is aesthetically pleasing. If your design meets all these requirements, you are ready for the next important step in the production of your illustration.

The final or comprehensive layout should be checked carefully with your client: scientist, author, publisher, or editor. It is best to perform this step personally with your client, who may have a different approach in mind. It is the responsibility of the artist to design and to produce the finest graphic communication of which he is capable. It is also his responsibility to explain the rationale of his method to his client.

If any alterations must be made in the drawings, design, or copy, they should be made now while you are still in the planning stages. Any later changes will be very costly for your client in terms of money and for you in terms of time and energy. Prepare the drawings with tissue overlays to indicate changes. After checking and making any corrections or alterations the client should initial the illustrations. You are now ready to render the final drawings and to order the type for the camera-ready copy.

9-33. a. Robin Ricks. Ink on scratchboard.

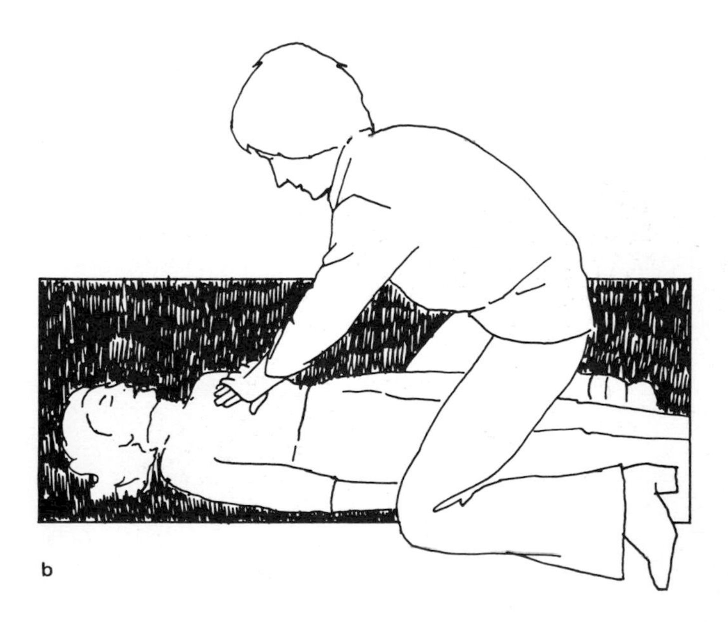

b

THE MECHANICAL

After the final or comprehensive layout (comp) has been approved by the client, the camera-ready copy, or mechanical, is drawn or mounted on white cardboard. All the type and illustrations are pasted in place or on registered overlays or are indicated on number-keyed illustrations.

Pasteup

All the type and line illustrations are cut out, arranged, and cemented directly to the board. Be as clean and neat as possible. When cutting out elements to be pasted up, use a straightedge, cutting parallel or at a right angle to the base line. An oblique cut creates an optical illusion, making the type or illustration appear crooked.

The elements can be cemented to the board with rubber cement, a glue stick, or melted and cooled wax from a waxing machine. A single coat of rubber cement is applied and quickly aligned before it dries. Most printers can wax the back of your type, and various styles of studio waxers are available. After adjusting the elements they should be burnished or rubbed down through a protective sheet of tracing paper.

Identification and printing information

Guidelines outline all the margins, keyed illustrations, and crop marks. They should be made in light blue nonreproducible pencil or pen so that they are filtered out by the camera. A T-square and a triangle are used to ensure that the lines are at right angles and that the margins are even.

The author's name and address, the artist's name, and the figure number are indicated on the mechanical. The top of the drawing or page layout is indicated by an arrow pointing up and with the word "TOP" at the top of the page (9-34). Reduction instructions should also be supplied.

Cover sheet

Protect the front of the mechanical with a paper cover sheet folded over and taped to the back. This also adds a professional appearance. Some artists consistently use a special color and quality of paper as their personal identity mark.

Halftones

There are two kinds of printed halftones, square and outline. Each has different advantages and different results. The artist should know early in the planning and rendering stages which method will be used to reproduce the drawings.

Square halftones

Most continuous-tone illustrations are printed as square halftones, as they are the easiest and least expensive to produce (9-35). Continuous-tone illustrations are standardly printed as square halftones unless special instructions are given to the printer.

Square halftones are prepared by cutting the opaque mask on the overlay in a rectangular shape covering the drawing. Every part of the rectangular area is screened. Even the white areas of the illustration register a light gray value as a result of the smallest dots on the screen. In this method there is a gradual tone change on the reproduction similar to that of the original.

Outline halftones

On an outline halftone, also called a silhouette or dropout halftone, only the gray and black values of the illustration are screened. The white parts of the illustration (background and highlights) are white on the print without any dot pattern.

In a silhouette halftone the artist controls the areas that are screened or not screened by cutting and placing a mask of opaquing film on a registered overlay over only the area that he wishes to be screened. White (unscreened) areas or highlights within the illustration can be produced by removing the opaquing film from these areas (9-36). This method of producing an outline halftone creates a sharp definition between the screened grays and the unscreened white portions of the illustration.

The photoengraver, by exposing the illustration twice, can produce a dropout halftone. The first exposure records the grays and blacks. In this method the transition from screened grays to unscreened whites is more gradual than with the opaquing method and more similar to the original (9-37).

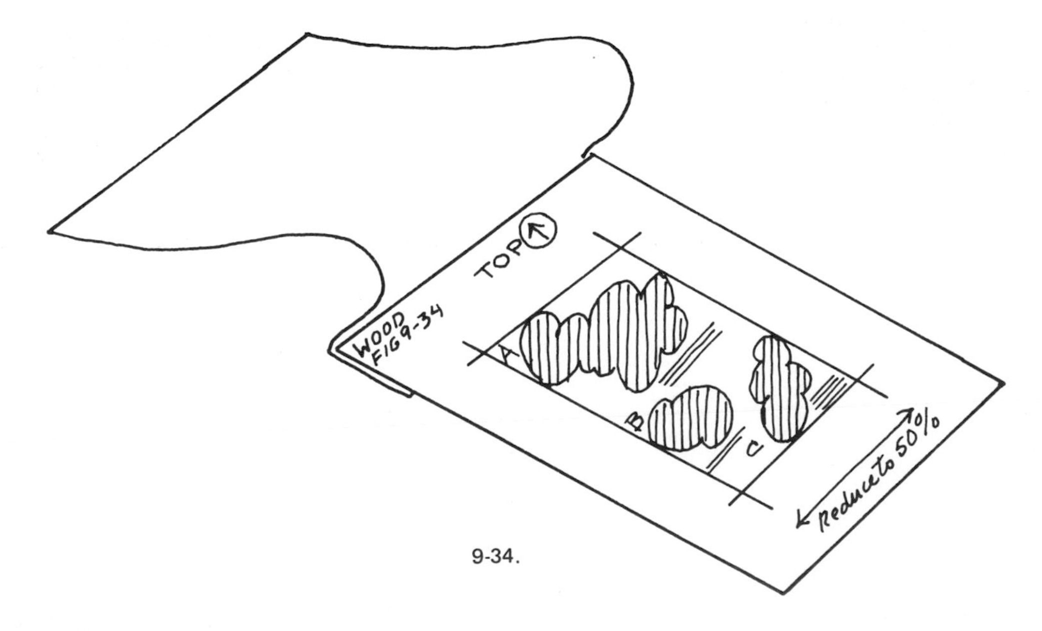

9-34.

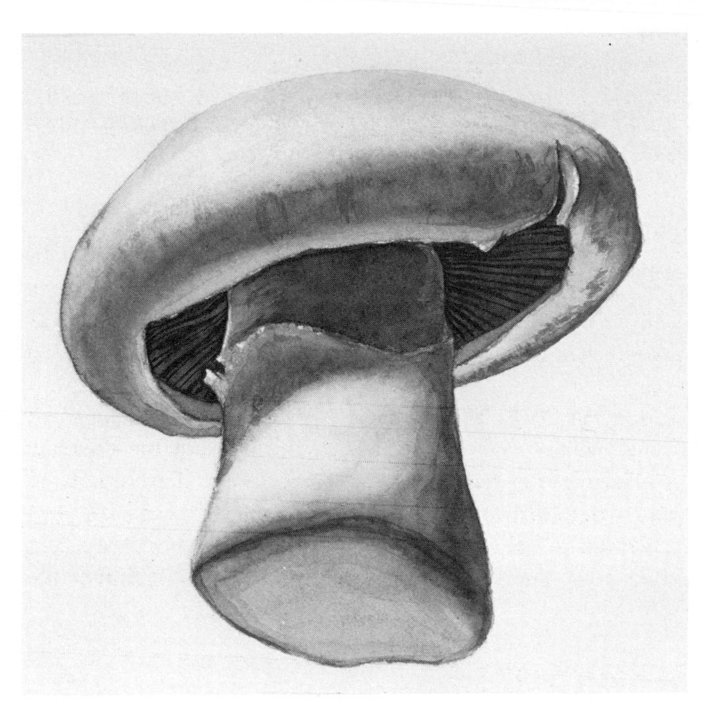

9-35. Anton Friis. Mushroom. Wash on cold-press illustration board.

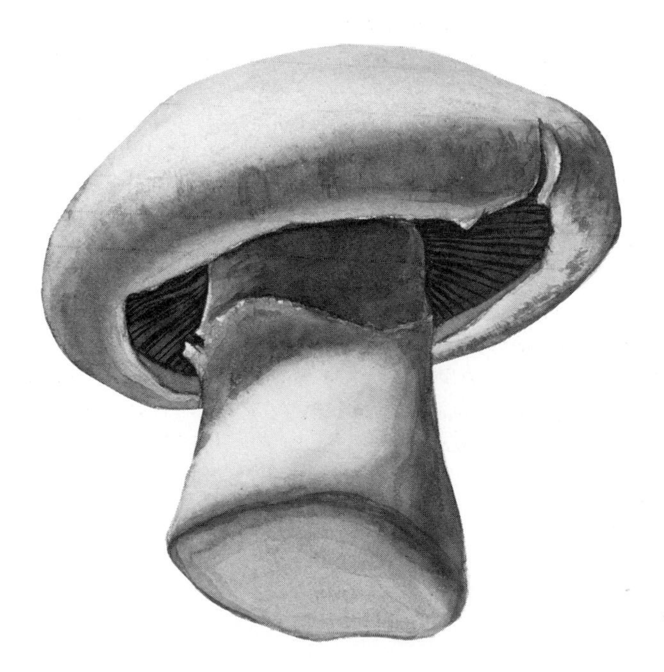

9-36. Anton Friis. Mushroom. Wash on cold-press illustration board.

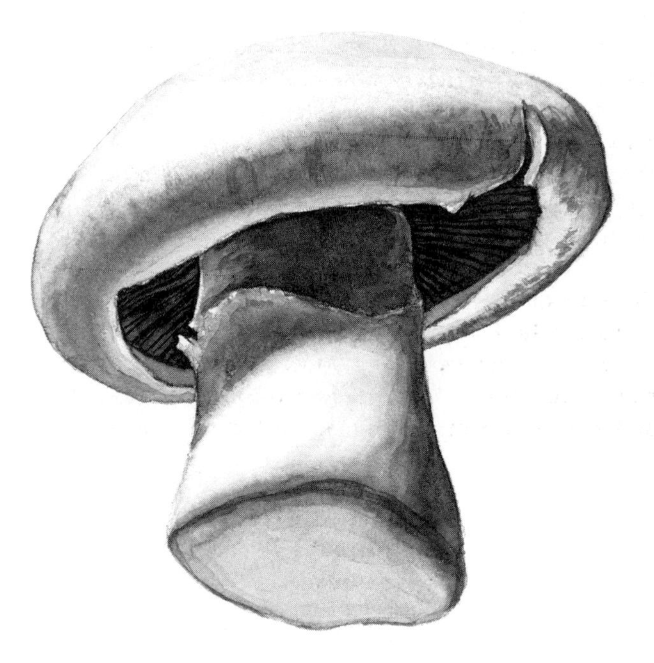

9-37. Anton Friis. Mushroom. Wash on cold-press illustration board.

Continuous tone plus line

Continuous-tone illustrations are treated differently from line copy by the photoengraver and must be arranged so that they can be photographed separately.

A tone illustration can be mounted on the board and three registration marks placed just outside the margins. All line illustrations, type, leaders, and corresponding registration marks are then mounted on a stable overlay (usually of mat acetate). If a silhouette halftone is desired, the overlay area on the tone illustration is covered with an opaque material, usually a dark red film that appears black to the camera. Some of the trade names are Rubylith, Transopaque, and Parapaque. There are also liquid opaquing materials. India ink may be used to touch up the edges. This opaque mask creates a window in the negative in which the screened continuous-tone copy is to be added. The photoengraver then makes a line negative of the line copy and the window and a screened negative of the continuous-tone drawing (9-38).

In another method all the copy, screened and lined, is included on the same board. The overlay is a transparent film registered to the board that contains the opaque mask over the material to be screened. This is a particularly good method when the tolerances between the line and screened elements are tight (9-39).

A third way to handle the combination of screened and line copy on the same plate is to put all the line copy on the board. Opaque material is affixed to this same board to produce the windows on the negative in which the screened tone illustrations are to be placed. The continuous-tone illustrations can then be placed on a registered overlay (9-40).

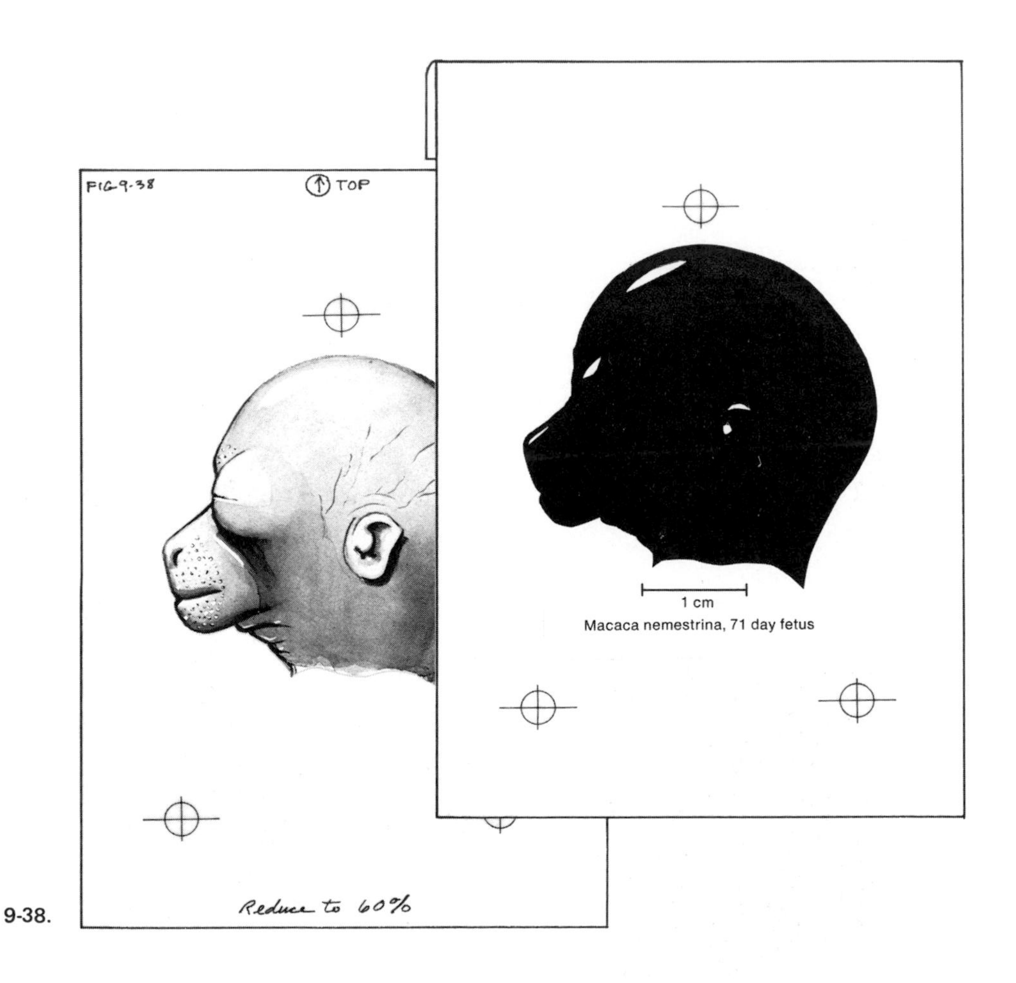

9-38.

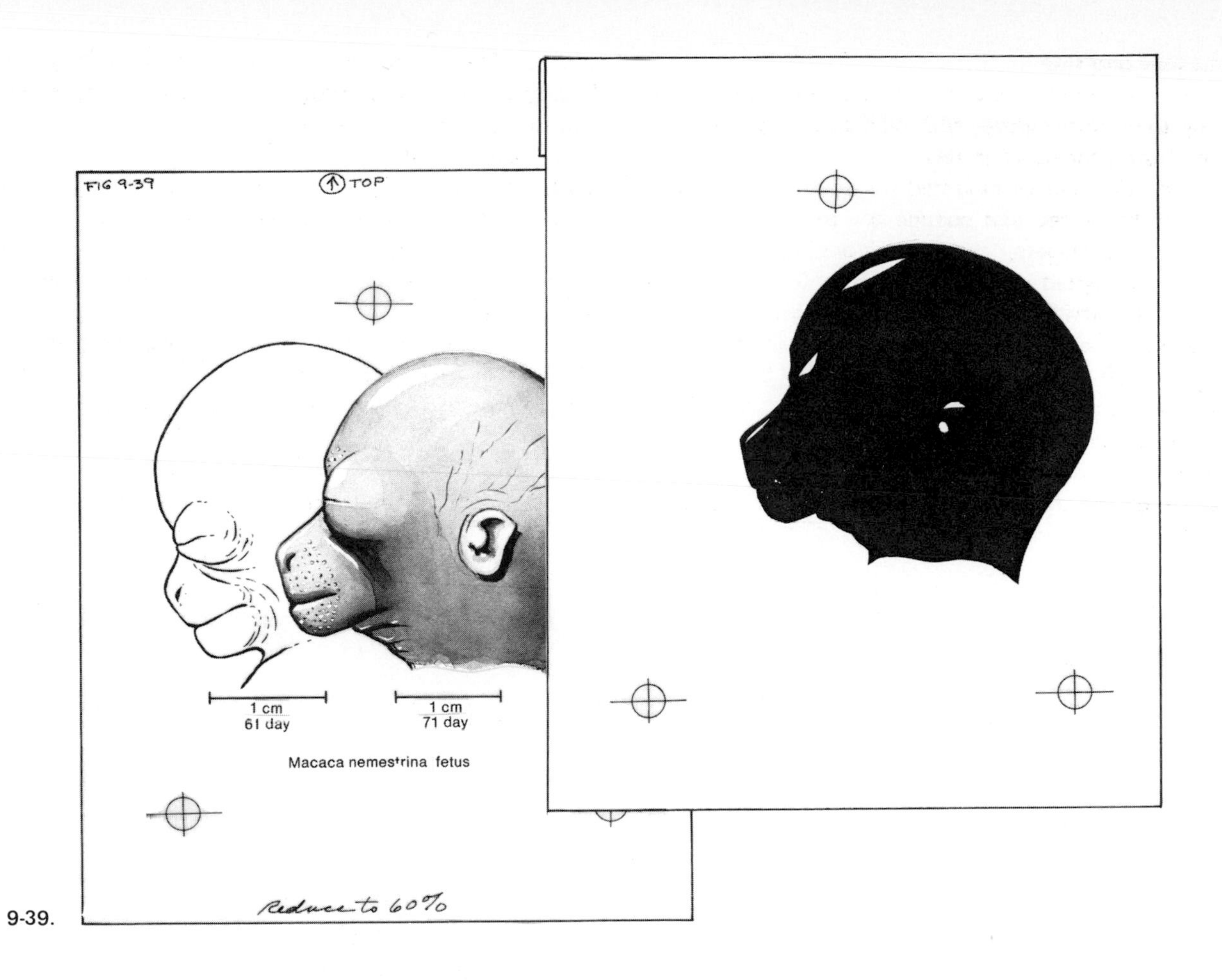
FIG 9-39
TOP
1 cm
61 day
1 cm
71 day
Macaca nemestrina fetus
Reduce to 60%

9-39.

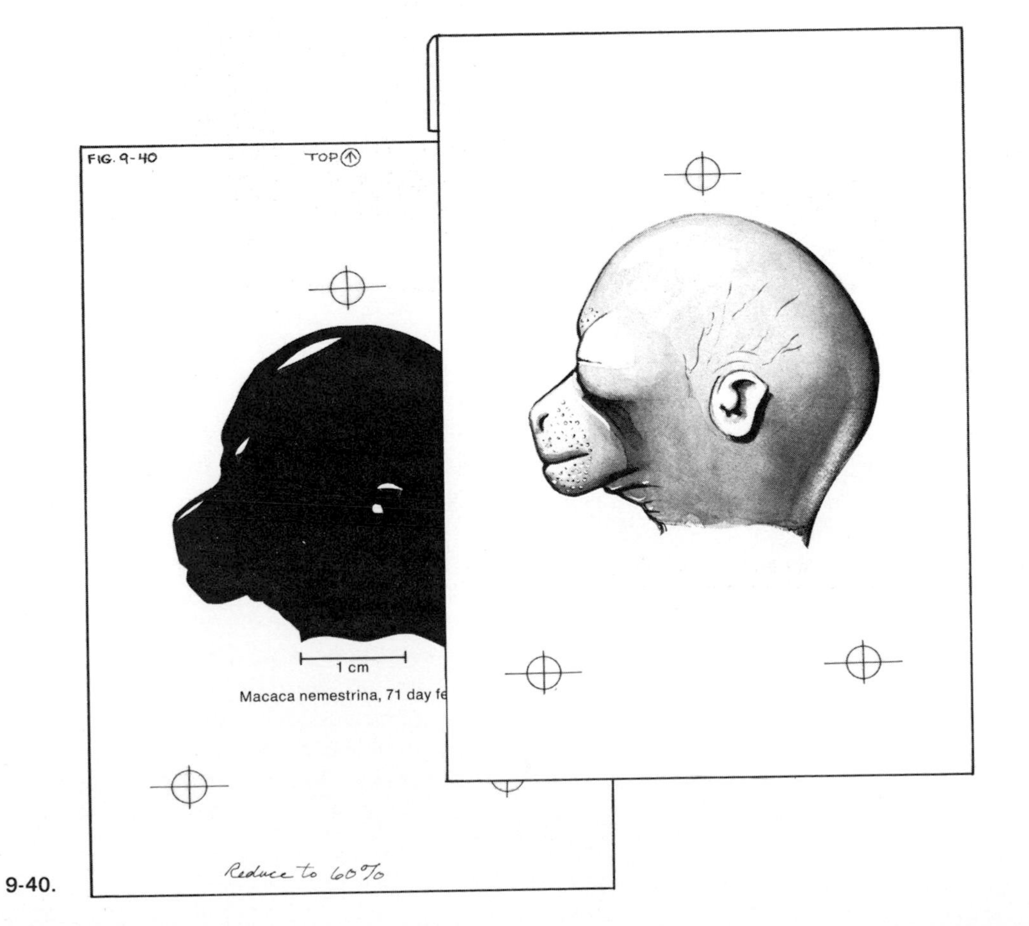
FIG. 9-40
TOP
1 cm
Macaca nemestrina, 71 day f
Reduce to 60%

9-40.

If the tone illustrations are to be reduced to a different percentage than the line copy, they are not mounted in place on the overlay but are keyed or numbered to correspond to the numbers on the mechanical. The percent of reduction is noted directly on the margin of the illustration. If the illustration is to be cropped, a paper frame is taped over the illustration, revealing only the area to be printed, or blue crop marks are placed in the margins.

Photographs

The illustrator is often required to organize, mount, and label photographs or to incorporate them into an illustration. Black-and-white or color photographs must be screened for publication and separated on the mechanical in the same manner as for any continuous-tone or color illustration.

Handling

In handling remember that the emulsion side of the photograph is delicate and can be easily damaged by fingerprints, bending, or pressure from either side. Protect the front of each photograph with a cover sheet that is folded over and taped to the back. Do not use paper clips on photographs. If you must write on the protective flap or on the back of a photograph, use a soft pencil or a felt-tip pen and a light touch. A gummed label on the back is preferable. Do not roll photographs, as the emulsion may crack. Keep them clean: a fragment rubbed or pressed into the emulsion may mar it. Any grooves or marks in the surface may be visible in the reproduction.

Mounting

Photographs can be sent to the publisher either loose or mounted. Mounting gives them added protection and allows room for identification and crop marks, but many publishers prefer loose photos. If there are a number of photographs that are to be printed in a certain sequence, you will be sure of the result if you mount them. White Strathmore 2- or 3-ply illustration board or other sturdy board is satisfactory for mounting.

Trim all the white borders. Make guidelines on the mounting board with a light blue nonreproducible pencil, using a T-square and a triangle to make sure that everything is lined up properly before mounting.

There are a number of methods of mounting photographs, which vary in durability. Dry mounting produces a flat, long-lasting bond. Dry-mounting tissue is trimmed flush with the photograph. A heated dry-mounting press or hand iron melts the adhesive in the tissue, forming a bond between photograph and mounting board. Rubber cement, applied only to the mounting surface, produces a satisfactory bond. When applied to both photograph and mounting surface, it is a very strong bond. Double Nothing (a sheet of rubber cement enclosed between two sheets of protective paper) makes a good permanent mounting adhesive. Spray glue comes in both strong- and light-bonding grades. Small, light photographs can be adhered with a glue stick. Waxing the back of a photograph with a waxing machine, which uses melted and cooled wax as the adhesive agent, is not considered permanent. Double-sided transparent tape placed inside the guidelines on the mounting board makes a moderately secure bond.

When mounting or ganging a number of photographs for a plate, butt them neatly up against each other, with no white space showing between them. On an overlay indicate to the engraver any hairlines to be scribed between the photographs. This gives a neat appearance and is the most economical form to print, requiring only one mask instead of one for each photograph. It wastes none of the page area, with the result that the photographs can be printed as large as possible (9-41).

Cropping

Photographs may be cropped by cutting them to the desired shape and size and then mounting them. If you do not wish to cut them, you can put crop marks in the borders of the photographs or on the mounting board (9-42). Another method is to make a paper overlay and to cut out a frame around the area to be printed (9-43). Instructions for the printed size would read, "reduce to 10 picas," "enlarge to 20 picas," or whatever.

Labeling

Labeling is usually done directly on the photograph, as this is the most economical reproduction method. Dry-transfer letters can be rubbed directly onto the surface of the photograph. Take care to rub gently, or you may damage the emulsion surface. Transfer letters that are cut out of film are not as satisfactory, because the outline of the cutout film may reproduce. Inked lettering with templates can also be done on the surface of the photograph. If the ink does not adhere smoothly, wipe the area with a bit of spit on cleansing tissue, let dry, and reink.

The lettering is screened along with the photograph, so the contrasts will not be quite so crisp as they are on the original. Black or white lettering, leaders, and arrows should be chosen for greatest value contrast. Arrows and leaders are available in a black-and-white combination to contrast with both light and dark values (9-41).

If the labeling is outside the area of the photograph, it is better to do it on a registered overlay. In this case all labels, leaders, and arrows will be shot in black line instead of screened.

Identification

All photographs, mounted or unmounted, must be identified completely on either the back or the front. Include the author's name and address and the figure number. Identify the top of the photograph by writing the word "top" next to an arrow pointing up at the top. If the photograph is not mounted on the mechanical, the figure number is keyed to the space allotted to it: "place fig. 10-6 here," for example.

9-41.

9-42. Kay Rodriguez. *Macaca nemestrina*, infant-twin pigtail monkeys.

9-43. Kay Rodriguez.

CHAPTER 10.

Printing for Publication

In order to prepare illustrations appropriately for printing, you should know what happens to your artwork from the time that it leaves your studio until it appears on the printed page. The client, the printer, and the artist all contribute to the decisions that guide this transition period. As a professional artist you should be aware of printing processes in order to plan these procedures, advise your client on the options available to him, or work within already finalized plans.

You need to know what type of publication your artwork will be a part of: a textbook or journal, a thesis or grant application, a booklet, a self-mailer, or a poster. You should know the size, weight, and finish of the paper on which it is to be printed. You should also be aware of the various printing and binding methods and the advantages and limitations of each.

If you are fortunate enough to be working with a local printing house, arrange to take a tour through the plant. Printers are most helpful in explaining their equipment and procedures to an artist or prospective client. The printer is your most valuable source of information during all stages of planning. Each printing operation has different capabilities and specialties. The printer who produces the finest halftones and long-print books is not necessarily the best printer of a short-run booklet with line cuts.

PRINCIPLES

Each printing method has cost-effective characteristics that vary with the requirements of the job, depending on aesthetic qualities, time factors, and economic limits. These must be understood and balanced in order to select the correct method for the job.

Black ink, white paper

The principle of applying areas of ink from a printing plate onto paper is basic to every printing method. In the line-cut method the black lines of the drawings and lettering on the mechanical are transferred to a metal printing plate and printed on the paper (10-1). This can be done in the same size as the mechanical, reduced, or enlarged. A drawing that is rendered in continuous tone (shades of gray done with wash, pencil, carbon dust, or airbrush) must be screened, or broken into tiny dots of black, in order to be printed in black ink on white paper (10-2). This type of printing, known as the halftone process, does not print with gray ink. It prints black dots that appear as different values of gray. These dots are not visible without a magnifying glass except on very coarse screens such as those used for newspaper photographs. The eye interprets the dots as different values of gray. The areas of small dots form the light gray values, and the areas of large dots form the dark values (10-3).

Halftone screen

A grid pattern or screen is used to produce the dots photographically. Grid patterns vary greatly as to the number of grid lines per square inch. A 100-line screen has 100 dots per linear inch (2.5 cm). This means that in a 100-line screen each square inch (6.5 cm^2) has 10,000 dots. Newspapers use grid patterns of 55-, 65-, or 85-line screens (10-4). A 120-, 133-, or 150-line screen is average for scientific publications (10-5). Special printing jobs may have 200- or 300-line screens (10-6). The finer the screen, the more dots per square inch. The more dots per square inch, the truer the printed halftone to the original continuous-tone drawing.

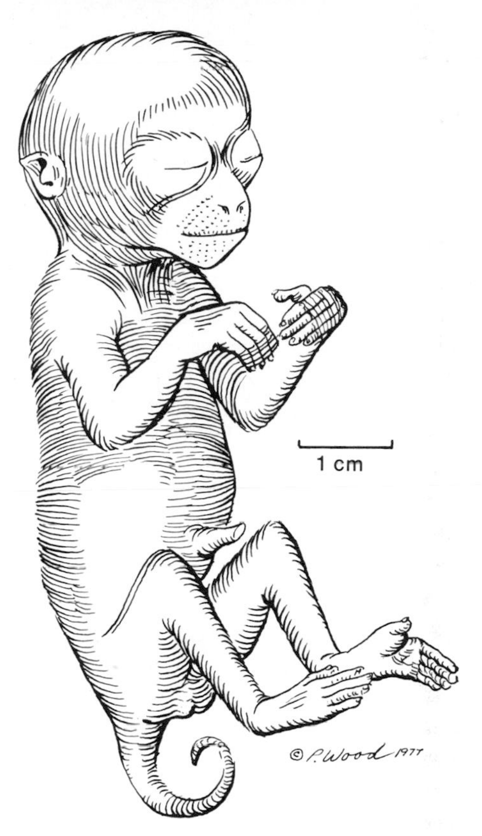

Macaca nemestrina, 71 day fetus

10-1.

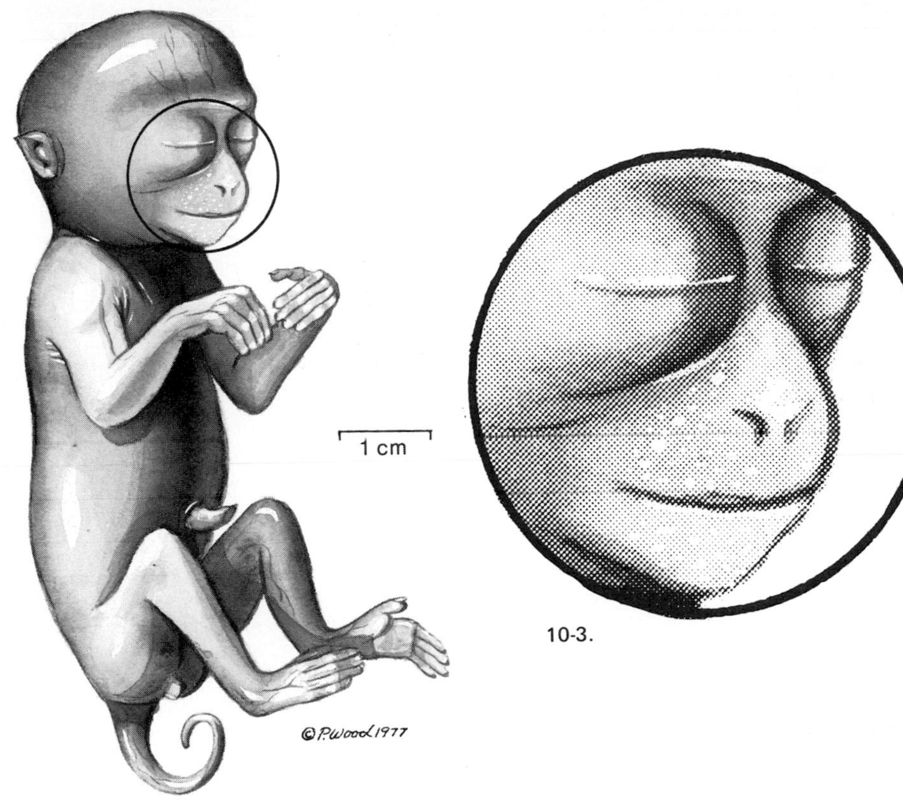

Macaca nemestrina, 71 day fetus

10-2.

10-3.

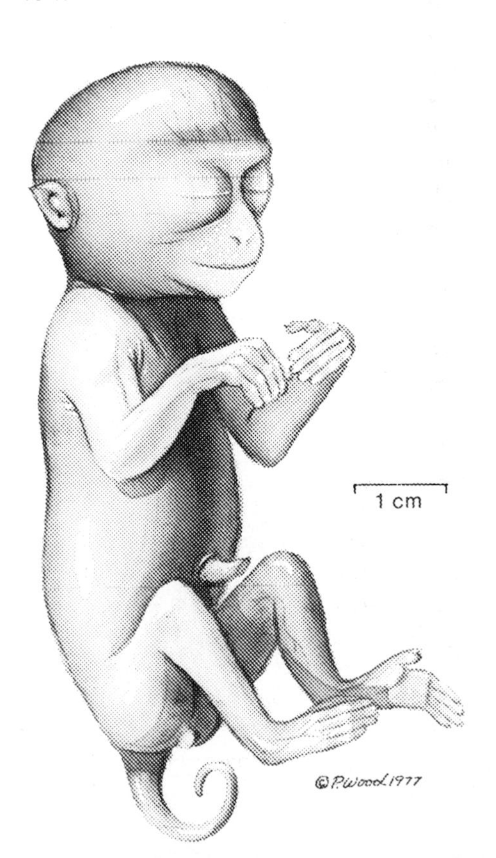

Macaca nemestrina, 71 day fetus

10-4.

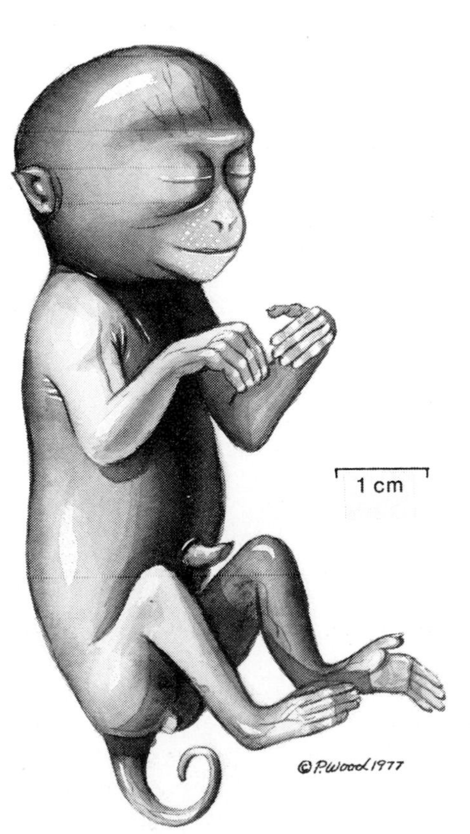

Macaca nemestrina, 71 day fetus

10-5.

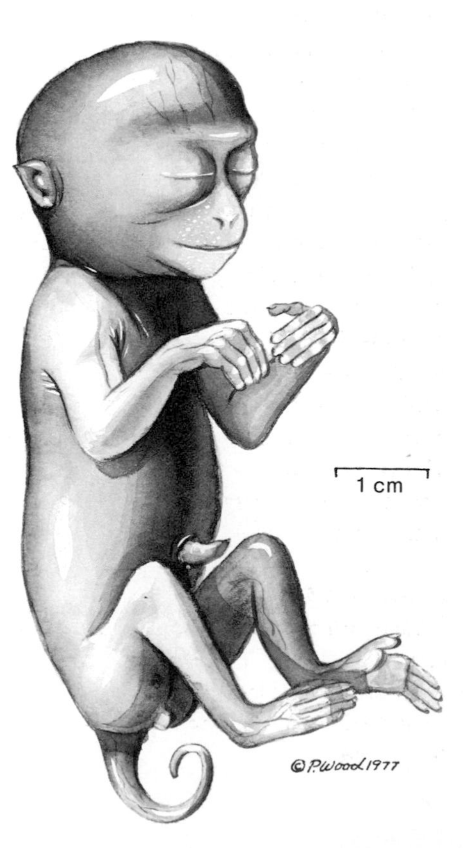

Macaca nemestrina, 71 day fetus

10-6.

Methods

Ink is transferred from the printing plate to the paper by one of four printing methods: planigraphic, intaglio, relief, and screen printing.

Planographic or offset printing

"Planographic" refers mainly to offset-lithography printing, which is generally called "offset" (10-7). The first lithograph plates were smooth or *plane* stones that were treated by hand to be ink-receptive and water-repellent on the areas to be printed black. The white or unprinted areas were the opposite, water-receptive and ink-repellent. This process is now done mechanically on a metal plate from negatives. The metal printing plate is flat, without indentations or raised areas. The term "offset" refers to the transfer of the ink to a second surface (rubber blanket) before it is transferred (or offset) to the final paper.

The transfer onto the smooth rubber blanket produces crisp, smooth edges and clean, clear printing on a wide variety of paper textures from rough to smooth. The mechanical can be transferred to the printing plate with a minimum of makeready time by the printer. This is a most economical and high-quality method, and most of your illustrations will be printed this way, as it combines copy and illustrations most satisfactorily.

Intaglio or gravure printing

In intaglio or gravure printing the surface of the printing plate has depressions or engravings that receive the thin ink (10-8). The excess ink is wiped off the smooth part of the plate by a wiper or doctor blade, and the ink from the indented areas is pressed onto the paper. The entire area of the plate must be screened—type and line drawings as well as halftones. The quality of halftone reproduction is excellent, but the platemaking cost is high, reserving the method for very long-run pieces. Postage stamps, paper currency, and the rotogravure section of the Sunday paper are familiar examples.

Relief or letterpress printing

Relief or letterpress printing is the only method in which the raised areas of the printing plate receive the ink and transfer it directly to the paper (10-9). This method is favored for printing text alone, but it is expensive for reproducing illustrations, as it takes a great deal of time for the printer to build the press form with equal pressures on light and dark areas.

Screen or silkscreen printing

Screen or silkscreen printing is not to be confused with the screening of a continuous-tone drawing into a halftone consisting of a gridwork of dots. In silkscreen a stencil is united to a fabric or metal screen through which a squeegee forces heavy, paintlike ink (10-10). This produces flat areas of color and not continuous-tone reproductions. The texture of the screen is recognizable on the printed area. Versatility is the principal advantage of screen printing. As any size, shape, or weight of material can be printed on, this is a good method to use for exhibit panels and posters.

Reprography or copying machines

There are many types of copying machines in use today; the most useful are electrostatic copiers and offset duplicators.

Electrophotography

In electrophotography selenium or zinc oxide, which holds an electrostatic charge, is used to transfer a black image to the copy paper. Xerography is one example of this process. These machines vary in quality, depending on the brand and their care and condition. Pencil and ink line reproduce well, but large areas of black fade. Enlargement or reduction is possible on some models, and color can be copied on others. This method of copying has an economical breakeven point below 10 to 50 copies.

Offset duplicating

Offset duplicating is done on a small lithographic press. The printing plate is paper instead of metal, necessitating a new plate after a certain number of copies are made. Some machines print in color and can use color or heavy-stock paper. Paper sizes can range up to 14″ X 20″ (36 cm X 51 cm). The convenience, speed, and cost factor make this a desirable choice for moderately good-quality printing under 10,000 copies. Black-and-white copy must be camera-ready with no overlays, as the plate is made directly from the copy and does not go through the negative stage, as in offset job printing. Screened illustrations (veloxes) have variable results, depending on the machine and the operation. Reduction and enlargement are possible on some machines.

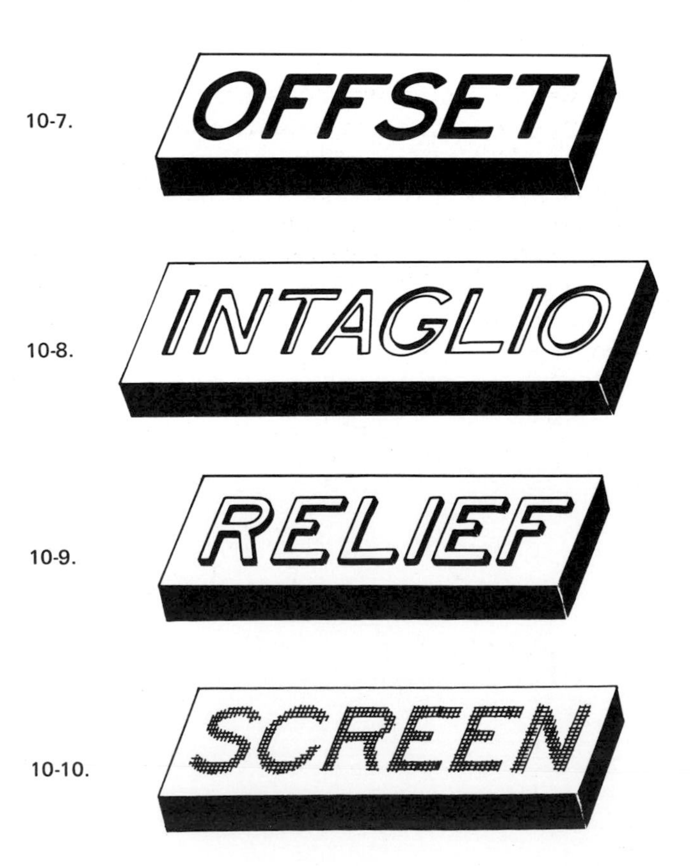

10-7.

10-8.

10-9.

10-10.

PAPER

The printed piece consists solely of paper and ink. The impression it creates therefore depends a good deal on the selection of the appropriate paper. Paper finish will affect the quality of the printing, whether line cut, halftone, or process color. Paper size, weight, and finish must be relevant to cost effectiveness. Paper color is a subliminal influence: it can be attractive, neutral, or irritating.

Your printer has certain basic stocks of paper on hand. Representatives of many paper companies can be helpful in advising you on the choice of paper and can give you samples for presentation layouts. If paper must be ordered, an additional time factor must be considered. Paper is usually ordered in large amounts, so for smaller jobs it is more economical to select from stock on hand. If the size and quality of paper are already selected by the client and publisher, the illustrations must be planned in a compatible size, technique, and amount of detail.

Size

Paper usually consumes 25% to 50% of the cost of a printed job, so it is important to avoid waste in planning the size.

The illustrator is primarily concerned with two paper sizes: text (or book) and cover. Text papers measure 25″ X 38″ (64 cm X 97 cm). This size divides into eight pieces measuring 9″ X 12″ (23 cm X 30 cm), including trim on the edges. It divides into only four pieces measuring 11″ X 14″ (28 cm X 35 cm). Cover paper, which is heavier, measures 20″ X 26″ (51 cm X 66 cm). Most cover stock can be purchased in double size, 26″ X 40″ (66 cm X 101 cm).

Weight

Paper is described by weight. The weight is computed by weighing 500 sheets (1 ream) in the standard size: 500 sheets of text paper (25″ X 38″ or 64 cm X 97 cm) weighing 80 pounds would be described as 80-pound paper or basis 80; 500 sheets of cover paper (20″ X 26″ or 51 cm X 66 cm) weighing 80 pounds would also be described as 80-pound paper or basis 80. As text paper is almost twice the size of cover paper, 80-pound cover paper would be almost twice as heavy as 80-pound text paper.

Text papers range from light 35-pound stock to heavy 120-pound stock. Cover papers go from 50-pound to 100-pound stock. The same paper may come in various weights: with other variables equal, the heavier stock is more expensive.

In the metric system the weight is referred to as grammage and measures weight per square meter (g/m^2). The weight is therefore independent of the paper size and reflects the true weight of the paper stock.

The choice of the appropriate weight of paper reflects the image of the publication, the "feel," the strength, and the opacity or see-through quality of the piece. Heavier is not necessarily better. Examine and feel various printed papers to become aware of the characteristics of the various weights.

Finish

Paper finishes range from rough antique through smoother eggshell, vellum, machine finish, and English finish to very smooth supercalendered papers. Coated papers are even smoother. Text papers come in a variety of textures and colors, making them ideal for specialty pieces such as flyers, booklets, and brochures. The finish of the paper, along with the weight, determines the quality of the reproduction and the character of the piece.

Offset-printing quality is good on all textures and weights of paper except those made for letterpress only due to transfer on the smooth rubber blanket, but smoother papers produce a truer image with sharper definition in fine halftone screens and four-color process printing. Coated papers are even smoother, providing more ink-receptive surface and greater opacity.

Grain

All paper has a grain that is related to the manufacturing process. To find the grain, tear a piece of paper: it tears straightest with the grain (10-11). Paper should be folded with the grain for letterfolds, bookbinding, and booklets. A fold with the grain is smoother; the pages lie flatter; and the vertical pages (with the grain) are stronger. The underlined number in the paper size indicates with the grain: i.e., in paper size 25 X <u>38</u> the "38" is with the grain.

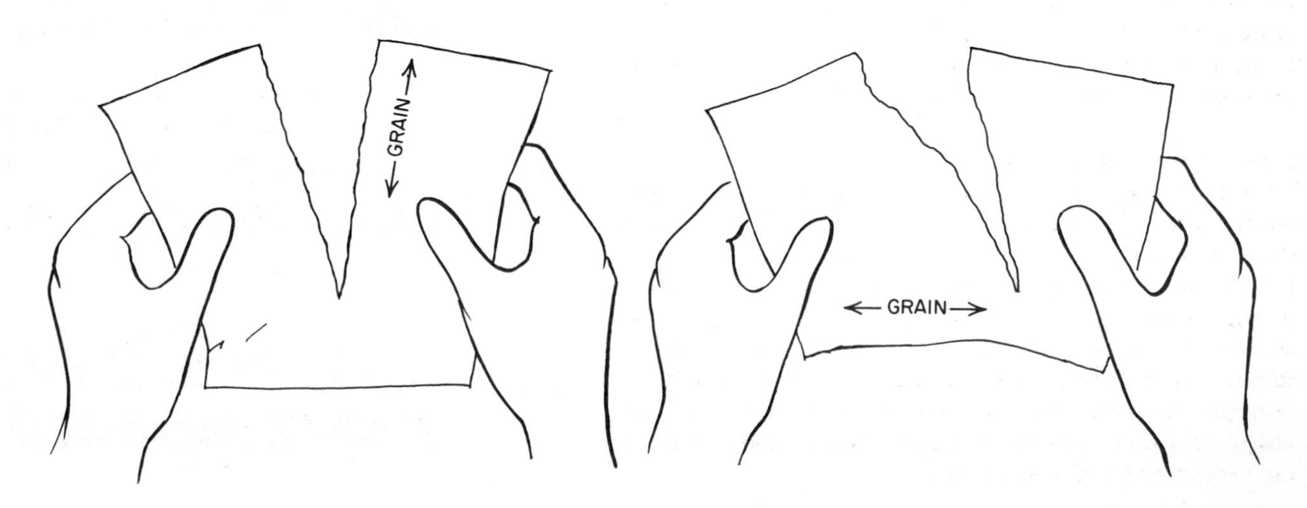

10-11.

FOLDING AND BINDING

Nearly every printed piece must be either folded or signatured and bound. This is the last step in the production and must be incorporated as part of the overall design plan.

Folding

Some printed pieces are cut to size and then folded as a finished piece. These include folders, leaflets, french folds, self-mailers, and maps. Folding is done by machine. Make sure that your printer has the equipment to fold your piece as you have designed it (10-12). This may save a costly hand operation. Heavy or chrome-coated papers must be scored before folding. You should be aware, while still in the design stage, that any ink on the fold lines might crack.

Signatures

Any piece that is to be bound is printed with several pages on a single sheet of paper in a prearranged order. This is called a signature. This sheet of paper is folded before being bound. The folding is done in a specified way so that the pages are arranged sequentially (10-13). One signature may be bound as a complete booklet, or many signatures may be gathered together in the proper order (collated) and bound.

Binding

The type of binding depends on the thickness of the publication, its life expectancy, and the desired quality.

Saddle stitching

Thin publications with either a self-cover or a paper cover of heavier stock are usually saddle-stitched. Saddle stitching consists of stapling (or stitching) several times in the gutter or fold of the publication. This allows it to be opened up flat. It is the least expensive method of binding (10-14).

Side stitching

If the publication is so thick that saddle stitching would create a fat roll at the binding edge or if it is beyond the capacity of the stitching heads, it can be side-stitched. In side stitching the sections or signatures of the publication are collated, or stacked in order. They are then stitched or stapled from the top to the bottom about 1/4″ (6 mm) from the binding edge. This measurement must be allowed for when planning the inside margins. The book may have a glued-on cover. It cannot be opened flat (10-15).

Perfect binding

Perfect binding is done on thick or medium volumes that are subject to ordinary use. The sections or signatures are collated as for a side-stitched volume; the back is ground off; and a flexible cover is glued on. The telephone book and most paperback books are bound in this way. They can be opened up flat (10-16).

Edition binding

Edition binding is the longest-lasting method of book binding. It is also the most expensive. Each signature or section of pages is sewn together with thread in the gutter or fold in a similar manner to saddle stitching. The signatures are then collated, glued at the binding edge, rolled for a round back, and covered with backing flannel; a hard cover is then affixed (10.17).

Mechanical binding

In mechanical binding each sheet of paper is separate. The sheets are punched with holes along the binding edge, and metal or plastic coils or rings are inserted. Some rings open to allow additional pages to be inserted. The cover is usually of a heavier stock than are the pages. The space for the coil must be accounted for in planning the inside or gutter margin. These books may be opened up flat, but they are not ideal for long, hard use (10-18).

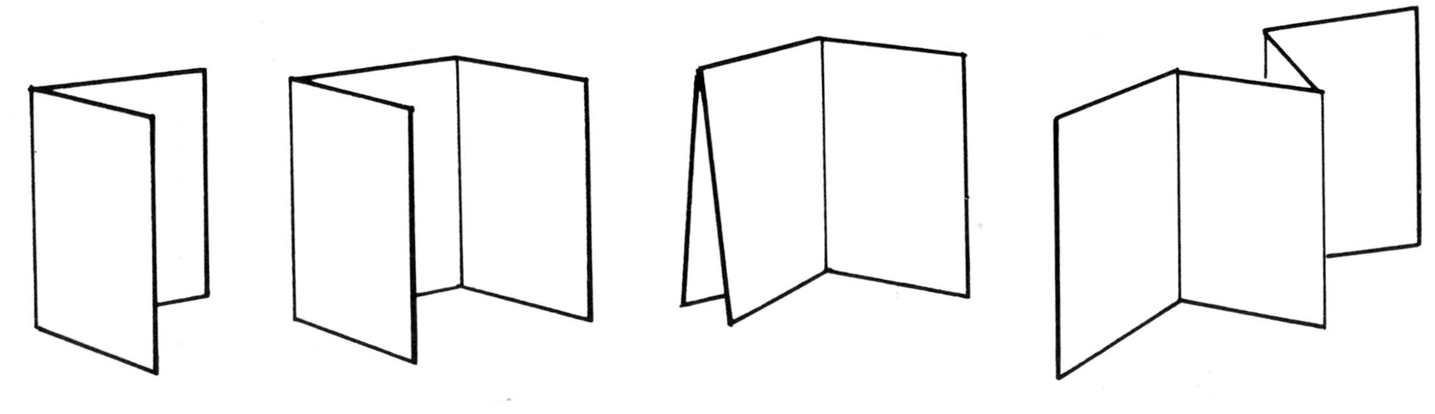

10-12.

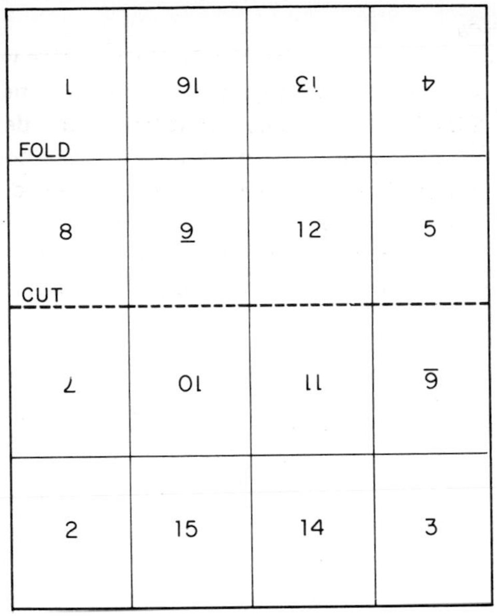

1	16	13	4
FOLD			
8	9	12	5
CUT			
7	10	11	6
2	15	14	3

10-13.

10-16.

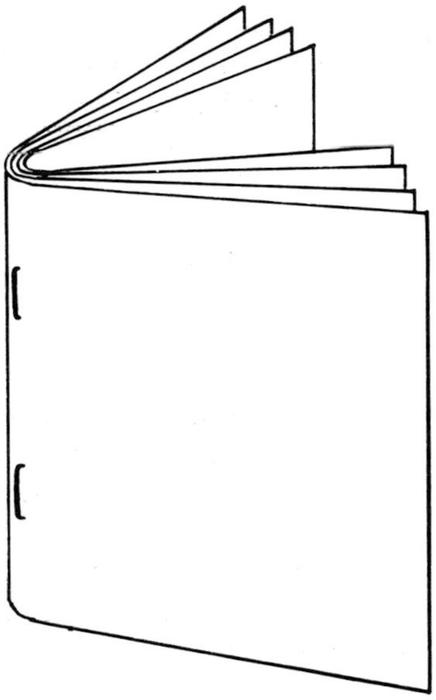

10-14.

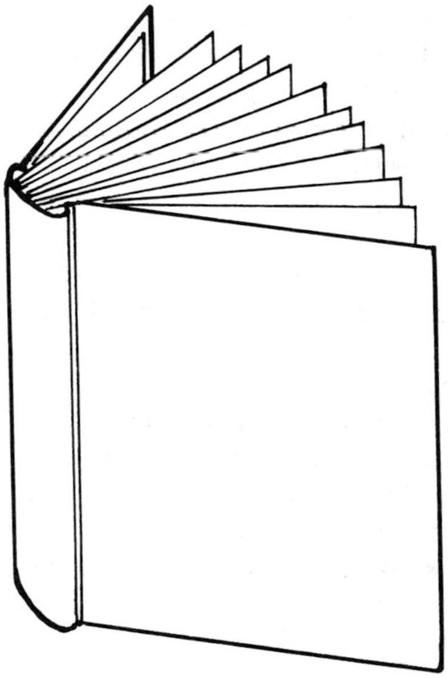

10-17.

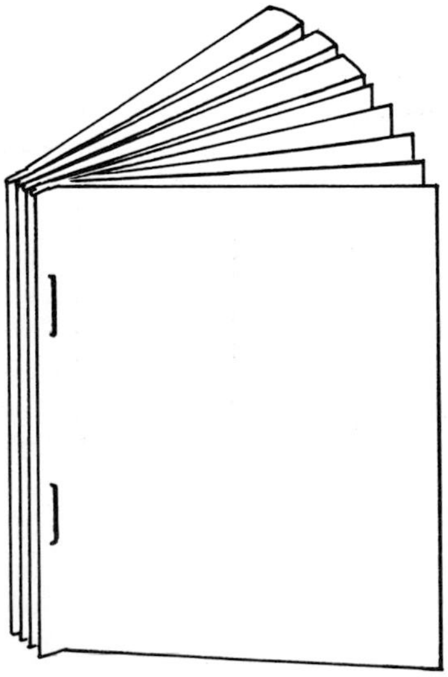

10-15.

10-18.

DEALING WITH THE PUBLISHER

The original illustrations, not photographic copies, should be sent to the publisher or platemaker. It is important that the platemaker have the very best material to work with, and an illustration is degraded to some degree each time that it is photographed.

Mailing

When mailing illustrations, package them carefully, as there is a possibility of rough handling in transit. They should be fully insured, with a return receipt requested. The amount of insurance should cover the entire expense of replacement.

Copies

Any illustrations that leave your possession should first be mechanically copied. Photographs or photocopies can be made if the expense warrants it. It is usually sufficient to duplicate them on a copy machine, even though this process distorts them slightly. If you must redo the illustration due to loss of the original, this is very helpful. It is also a point of reference in discussing alterations by phone or mail. The copies are useful in recalling the style and format used for this particular client or publication. A complete file of all your illustrations is also a great help in researching future projects.

Identification

Each illustration should be identified on the back with the name, address, and phone number of the artist and the author. A request for the return of the original drawing should be on the back of each illustration as well as in a separate letter. It may be returned by prearrangement to either the author or the artist. You can expect illustrations to be returned after the publication is printed.

Galley proofs

Galley proofs are sent to the author or the artist before publication. They are test prints of the pages, run on a hand press from the working plates, to allow a final check for errors. The artist is responsible for examining them carefully to see whether his instructions have been followed and whether the results are as expected. As the plates are already made, any changes (aside from printer's errors) are expensive, but any grave errors, such as an upside-down illustration, a smudged screen, or an unreadably reduced drawing, can be corrected at this last opportunity. If your instructions have been followed and the result is not satisfactory, part of the expense of making a new plate may have to be borne by the author.

Storage and retrieval

When the illustrations have been printed and are returned to you, they should be carefully boxed and stored so that they can be retrieved easily. With permission from the copyright holder they may be reused, or you may be asked years later to revise them for another edition of the same publication.

Copyright

Upon the release of a publication the publisher holds the copyright for the entire publication. This includes ownership and rights to reuse your illustrations. If you wish to retain the rights to future use of your illustrations so that you may use them yourself, control their reuse, or be paid for their reuse, you may release them for FIRST PUBLICATION RIGHTS ONLY. This may be written anywhere on the back or front of the illustration, along with your name, date, and the copyright symbol ©.

Reference or plagiarism

It is unethical and also illegal to trace, redraw, or revise an illustration without permission from the copyright holder. Scientific publishers are usually very generous in allowing reuse privileges to another scientific publication. This permission must always be obtained in writing, and credit must be given to the original publication in the caption. If another artist's drawing is revised, the new artist may sign it, adding after his signature "after (original artist)." Artists quite naturally use other artists' illustrations as reference and research material. There may sometimes be a very fine ethical and legal line between reference and plagiarism. If there is any question of plagiarism when your illustration is compared to your reference material, get permission from the copyright holder before publication.

CHAPTER 11.

Exhibits

The purpose of the scientific exhibit is to attract an audience, to stimulate interest in a subject, and to provide further information once interest is aroused (11-1). The information in a permanent exhibit can be more detailed and complex than in a temporary or traveling exhibit, but it should still be an exhibit in concept, treating a subject in an abbreviated overview rather than in an in-depth or complete manner, as a book or article would do.

The principles of exhibit design covered in this chapter include only simple methods of assembly. This does not mean to imply an amateurish appearance or approach. Easy-to-use materials handled in a straightforward manner and good design executed cleanly and neatly can stand proudly next to a commercially constructed exhibit.

PLANNING

In the first planning session the client and the designer or artist should discuss the following areas and reach at least partial decisions:

1. Schedule
2. Space
3. Budget
4. Audience
5. Information (words and pictures)
6. Slides and video
7. Size and shape
8. Printed material for distribution
9. Packaging and transportation
10. Assembling, dismantling, and storing

Schedule

Most exhibits are planned for specific meeting dates, leaving no leeway for the designer. For an exhibit of any complexity the designer should expect to need two months for planning and execution. A simple exhibit may need only one month to construct. Definite dates should be scheduled at the first conference for all subsequent client-artist conferences. There should be at least five conferences:

1. Preliminary planning
2. Rough layout completed—examination, proof copy, revisions, approval, budget
3. 6:1 scale model in color completed (6″ of exhibit = 1″ of model)—examination, revisions, approval
4. Full-scale (1:1) layout on butcher paper rendered in felt markers and taped to the wall at actual viewing height—final examination, proof copy, revisions, approval
5. Exhibit completed—demonstration of assembling, dismantling, and packing the finished exhibit

While this five-step process may seem repetitive, each step signals different problems, and various revisions, deletions, additions, and adjustments become apparent at subsequent phases of the process.

11-1.

Space

The exhibit designer must work within the restrictions of the physical space assigned to his client. This includes floor space, back and side walls, ceiling height, and existing lighting.

Exhibit areas at scientific meetings are usually divided into booths 6′ deep and 10′ or 15′ wide. They ordinarily have an 8′-high back drape and 4′-high side drapes. Tables, chairs, and additional lighting are usually available from a local display service. The exhibitor is assigned this space for the duration of the meeting, setting up the exhibit before the meeting starts and dismantling it on the last day of the meeting.

Some scientific meetings have poster sessions. Each exhibitor is assigned a bulletin board with an area usually 6′ to 8′ wide and 4′ high to which he attaches his material. The exhibitor stands by his exhibit during his allotted time, usually 1 to 4 hours, explaining his subject to interested colleagues.

Permanent in-house exhibit space may consist of glassed-in cases, channel molding for the insertion of flat exhibit panels, ceilings with screw eyes or a grid system to which hanging attachments can be made, bulletin boards, and other available floor or wall space (11-2).

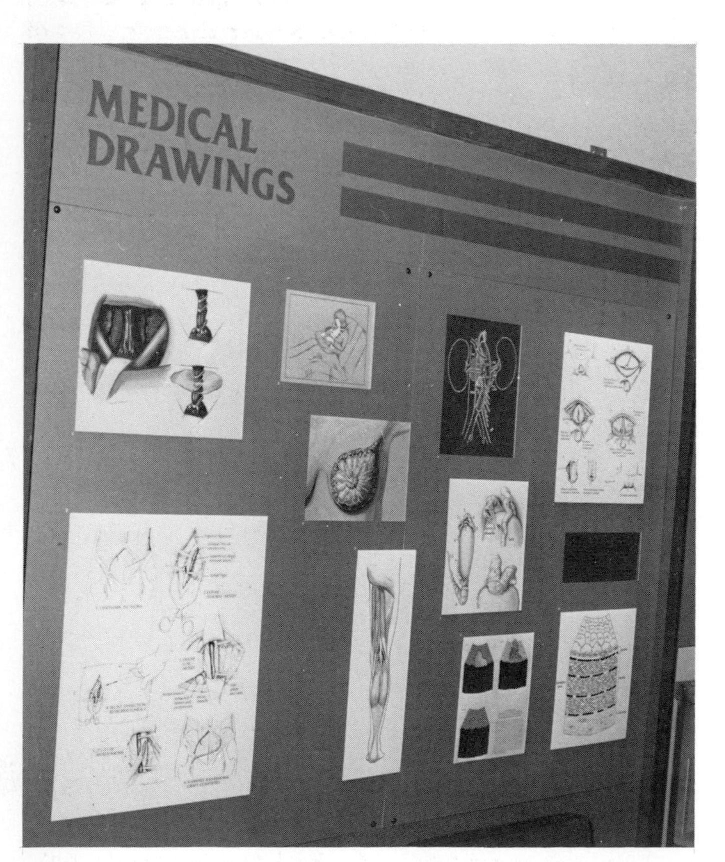

11-2.

Budget

The budget for the proposed exhibit should be discussed in the first planning session. The client's mental picture of his exhibit may have no relation to the money that he has to spend, and it is up to the designer to coordinate the client's dream with reality. If his dream is very strong, he may find that he can budget more to the exhibit; he may decide that he would be happy with a scaled-down version of his dream; he may decide that he must have *it* or nothing and would rather have nothing. In any case it is up to the designer to be realistic about costs. As in building a house planned expenditures tend to expand during construction, so it is wise to add a 10% contingency to the estimate for expanded costs.

The client usually has a specified amount of money available for the exhibit. It may include travel and expenses to and from the meeting, rental of the exhibit space and furnishings, and transportation of the exhibit in addition to the design and construction of the exhibit itself. Your budget should include time and materials. Following are time elements that you should allow for:

1. Conferences
2. Design
3. Layout
4. Production (art, photography, typography, etc.)
5. Revisions, additions
6. Transportation (to subcontractors—painters, silkscreeners, carpenters; to purchase special materials, tools)
7. Demonstration of assembling, dismantling, and packing

Following are materials that you should allow for:

1. Panels
2. Connectors
3. Paint
4. Lettering
5. Photographs
6. Carrying case

Keep accurate records of all your time and expenditures and notes on the size and complexity of the exhibit. This information, coupled with a documentary photograph of the exhibit, can help you in planning the budget for the next exhibit and also give your client a good idea of the results that he can expect. After several experiences in designing and producing exhibits you will be able to estimate the cost in linear measurements for freestanding, tabletop, and flat-hung exhibits of varying complexity.

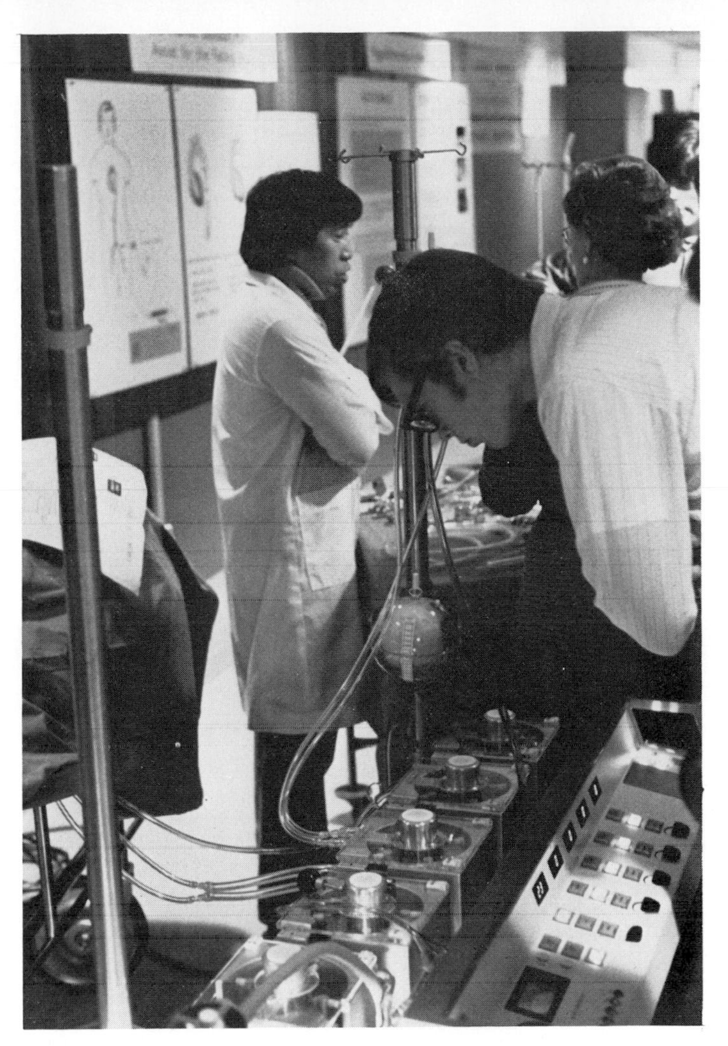

11-3.

Audience

The scope of the exhibit, including the written copy and the selected pictures, must be planned to reach the expected audience (11-3). An exhibit that is directed to the casual passerby, patient, or elementary student would have a different style, complexity, and level of information than an exhibit presented to one's colleagues. The presentation should not talk down to any audience of whatever age or educational level but should instead stimulate their curiosity and inform them in terms of their experience.

If the audience is expected to be large and crowded, more attention-getting devices may be necessary than for a small, select group.

Information

The information consists of the pictures—photographs and drawings—and the copy—titles, labels, explanatory text, and credits. These two elements are combined so that the copy ties the pictures together and the pictures enhance the copy.

The copy should be simple, easy to read, and short. The main titles should be intriguing enough so that the viewers want to read the text, ask the person tending the exhibit more about it, or take the printed handouts home to study. Most of the copy can be written in telegraphic phrases rather than in complete sentences (11-4). Remember that the viewer has a minimum amount of time in which to absorb the message.

EXPLANATORY COPY	EASY TO READ COPY
Chill the blood in ice, and place it in 0.6 ml plastic centrifuge tubes and centrifuge it at 800 x g for 10 minutes at 5° C.	Chill blood in ice. Place in 0.6 ml plastic centrifuge tubes. Centrifuge at 800 x g for 10 min at 5° C.
Remove the plasma carefully and place the fixative (15 glutaraldehyde in 0.1 m sodium cacodylate buffer) in the tube over the buffy coat. The fixation time is 30 minutes at 5° C.	Remove plasma carefully. Place fixative (1% glutaraldehyde in 0.1 m sodium cacodylate buffer) in tube over buffy coat. Fixation time is 30 min at 5° C.
Remove the fixative, and carefully cut the tube above and below the fixed buffy coat, gently remove the plastic ring, and hold it in 0.2 m sodium cacodylate buffer with 7% sucrose at 5° C for 1 to 7 days.	Remove fixative. Carefully cut tube above and below fixed buffy coat. Gently remove plastic ring. Hold in 0.2 m sodium cacodylate buffer with 7% sucrose at 5° C for 1–7 days.

11-4.

The pictures can inform or set a mood. If color is not an important factor, black-and-white photographs often carry more dramatic impact and are less expensive. Screening and posterizing are effective ways to reproduce mood-setting photographs. Posterization produces a black-and-white print of a continuous-tone photograph: the whites and the light grays in the photograph are white in the print; the blacks and dark grays are black. Posterizing can be done to adjust for minimum or maximum black or white. It is often necessary to touch up the negative or print in order to retain the significance of the original photograph (11-5).

Photographs should be printed from original negatives, not from published prints. They should have a mat (dull) finish rather than a glossy surface in order to avoid the glare from ambient-light sources.

Slides and video

A continuous-loop videotape or an automatically timed slide show from a rear-screen projector, with or without sound, may add interest and information to your exhibit. This matter should be carefully evaluated for its attention-getting characteristics, informative values, and competition with surrounding lights and noise. Individual earphones may be used for concentrated listening. A darkened area may be necessary for viewing (11-6).

If you decide to use a video or slide screen, it should be included in the overall design and not be added as an afterthought. An early decision should be made on screen size, as it can vary a good deal. The depth and space for such equipment must be requested on the exhibit application as well as electrical outlets, current, and extension cords needed. If the construction of the exhibit allows, the machine may be an integral part of the structure; otherwise a table of the right size and height must be acquired.

11-5. James Sneddon. *Macaca nemestrina*, mother and infant.

Size and shape

After you know the space available, the information that you wish to convey, and your budgetary limitations, you can decide on the size and shape of the exhibit—flat-hung, table-top, or self-standing.

Size

Exhibit space is usually measured in linear feet from left to right. The actual space can be increased by bending the display panels in an accordion fashion (11-7).

There is a limit to the amount of information that can be fitted into an area. There is also a limit to the amount of space needed for a specific amount of information. Just as 10′ (linear) of information cannot satisfactorily be squeezed into 6′, 4′ of information cannot be spread attractively over 10′.

The optimum viewing height is at eye level (5′). The prime viewing area extends 2′ above and below this level. A strip of the exhibit 7′ above the floor to 3′ above the floor is thus the prime viewing area. If the material to be viewed is small and requires close attention, only 1′ above and below eye level is considered to be the prime viewing area.

11-6. Cheryl Vigna and Margaret Watson, designers.

Added to this strip is the title or header of the exhibit, which is meant to be viewed from a distance in order to attract viewers. It should be at the 7 1/2′ level in order to clear the heads of the audience (11-8).

Flat-hung exhibits

Flat-hung exhibits consist of information on panels made of hardboard, foam-core board, plastic, or pressed board (Masonite). They depend on an auxiliary support for attachment. Foam-core and cardboard panels can be hung easily, but heavier plastic and wood panels must be hung with more sturdy attachments.

Flat-hung exhibits can be attached to the wall with push-pins, map tacks, or double-sided foam-adhesive tape. If they have permanent or revolving status, metal channels can be attached directly to the wall so that they can be slipped in and out. They can be grommeted at the top and suspended from the ceiling with nylon fishing line, which is almost invisible. In this way both sides of the panels can be used, serving as eye catchers, space dividers, or in groups as informative panels. Look at the ceiling that you wish to use. Many buildings have exposed pipes to which the lines can be attached. Some ceilings have a stable metal grid that supports removable insulation panels, which can be lifted to accommodate the attachments. Screw eyes can be attached to some ceilings for the fishing line. Three-legged easels and movable or permanent bulletin boards may be available for holding your exhibit (11-9).

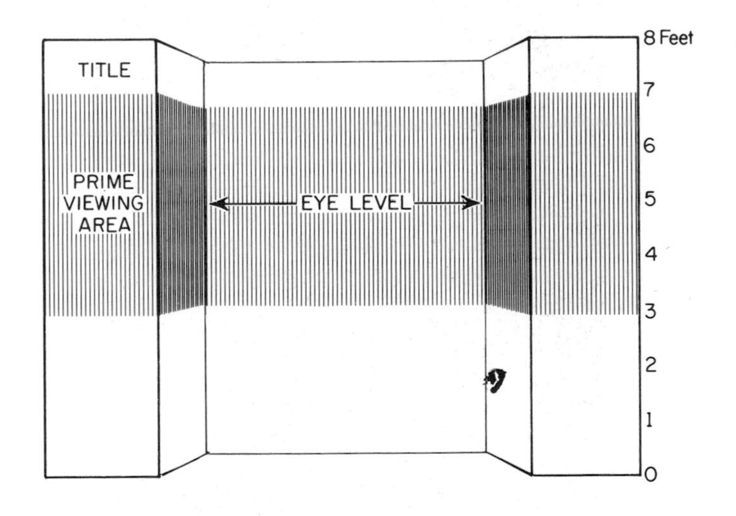

11-8.

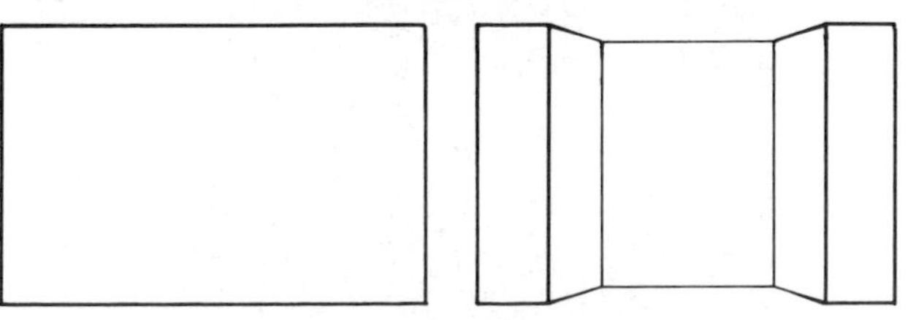

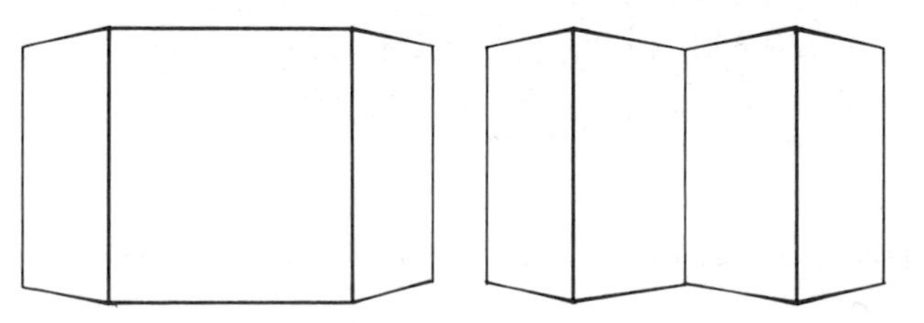

11-7.

Tabletop exhibits

Tabletop exhibits are constructed to be freestanding on a table, eliminating the difficulty of building the lower 3′ of the exhibit below the prime viewing area. Tables are rented in advance through most exhibit committees. The table legs are generally draped or skirted with paper, cardboard, or felt (Duvetyne) to harmonize with the exhibit color.

Ordinarily constructed of cardboard or foam-core board, tabletop exhibits are lightweight and easy to pack, assemble, transport, and disassemble. Demonstration materials can be placed on the table in front of the exhibit.

The simplest kind of support for single or double panels of foam core or cardboard is one or more easels attached to the back. The easel is glued or taped 1/4″ above the lower edge of the panel and cut at an angle so that the panel leans slightly backward (11-10). A long panel can be made secure by gluing a slightly shorter panel to its back and attaching either its lower edge or a glued-on L-shaped molding to the back edge of the table with C-clamps (11-11).

Panels can be scored through the outer skin with a blade and bent to form many freestanding configurations. Sets of boxes (fabricated, purchased, or found), each one slightly smaller than the next, can be covered with graphics and easily packed, each nesting inside the next (11-12). Panelocks are circular, clear-plastic slotted disks used to connect pieces of foam-core board at one of four angles (11-13).

Self-standing floor exhibits

The self-standing floor exhibit is considered the most versatile, handsome, and independent structure (11-14). It is also the most difficult to construct. Combining the elements of height, strength and stability, packing and transportation, maneuverable weight, and quick setup time is not an easy or inexpensive synthesis. Commercial exhibit companies build beautiful freestanding exhibits, but the cost is generally unjustifiably high for most scientific purposes (11-15).

Most tabletop configurations can be adjusted to freestanding floor exhibits, but a stable tabletop or model exhibit does not necessarily expand to a stable full-scale floor exhibit. The taller the exhibit, the more support it needs. A 90°- to 45°-angle support is necessary no less than every 3′ (1 m). The larger the individual panel, the more it tends to bend or warp. Flat lengths of wood glued to the backs of the panels help to support them. Even with this support an exhibit constructed of cardboard or foam core, because of its fragility, must be protected from viewer traffic, as it may easily be knocked down and broken. Stronger materials such as plywood or plastic can be built into a stronger, more stable permanent exhibit but pose weight and space problems in packing, transporting, and storing a traveling exhibit.

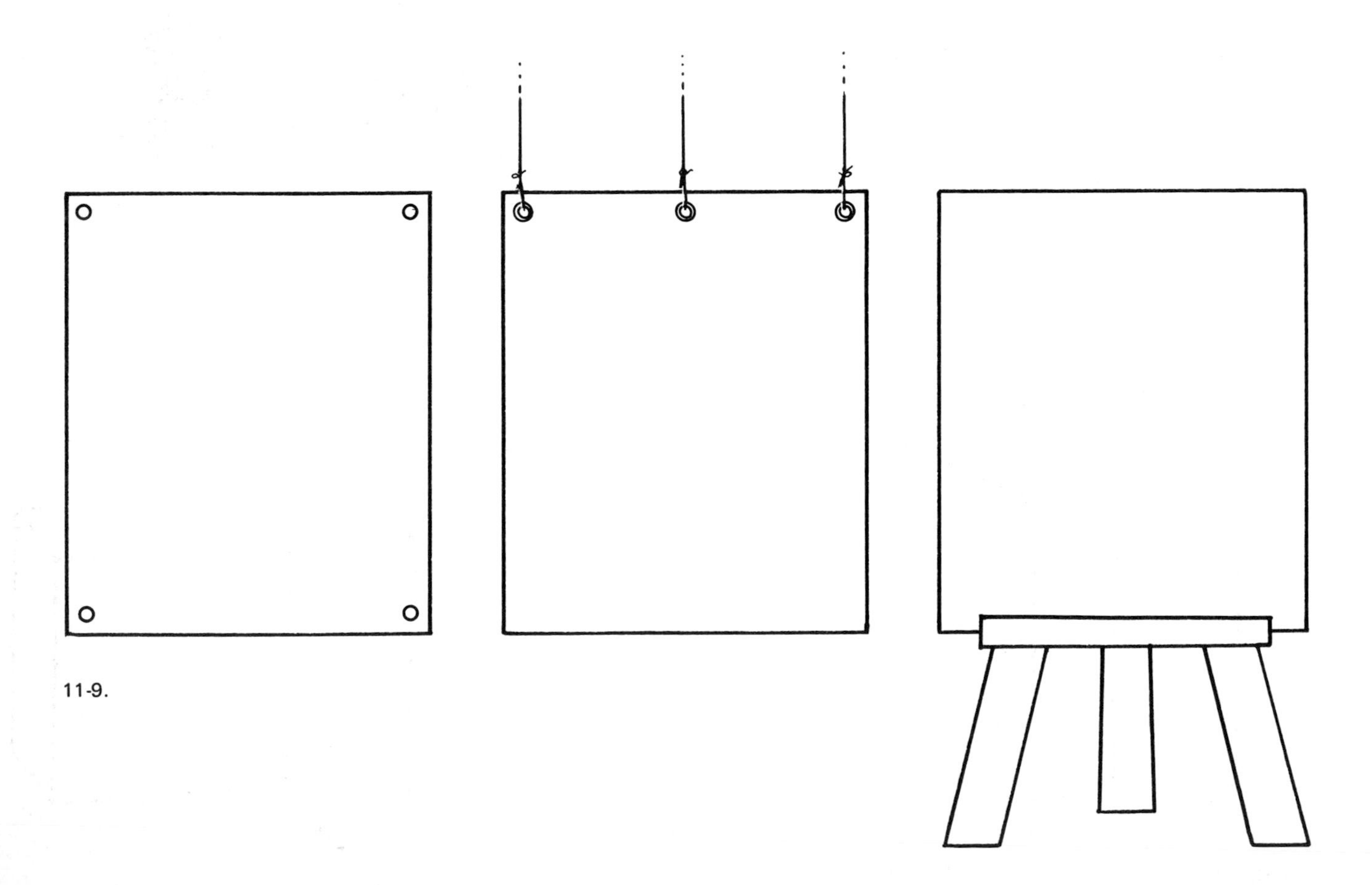

11-9.

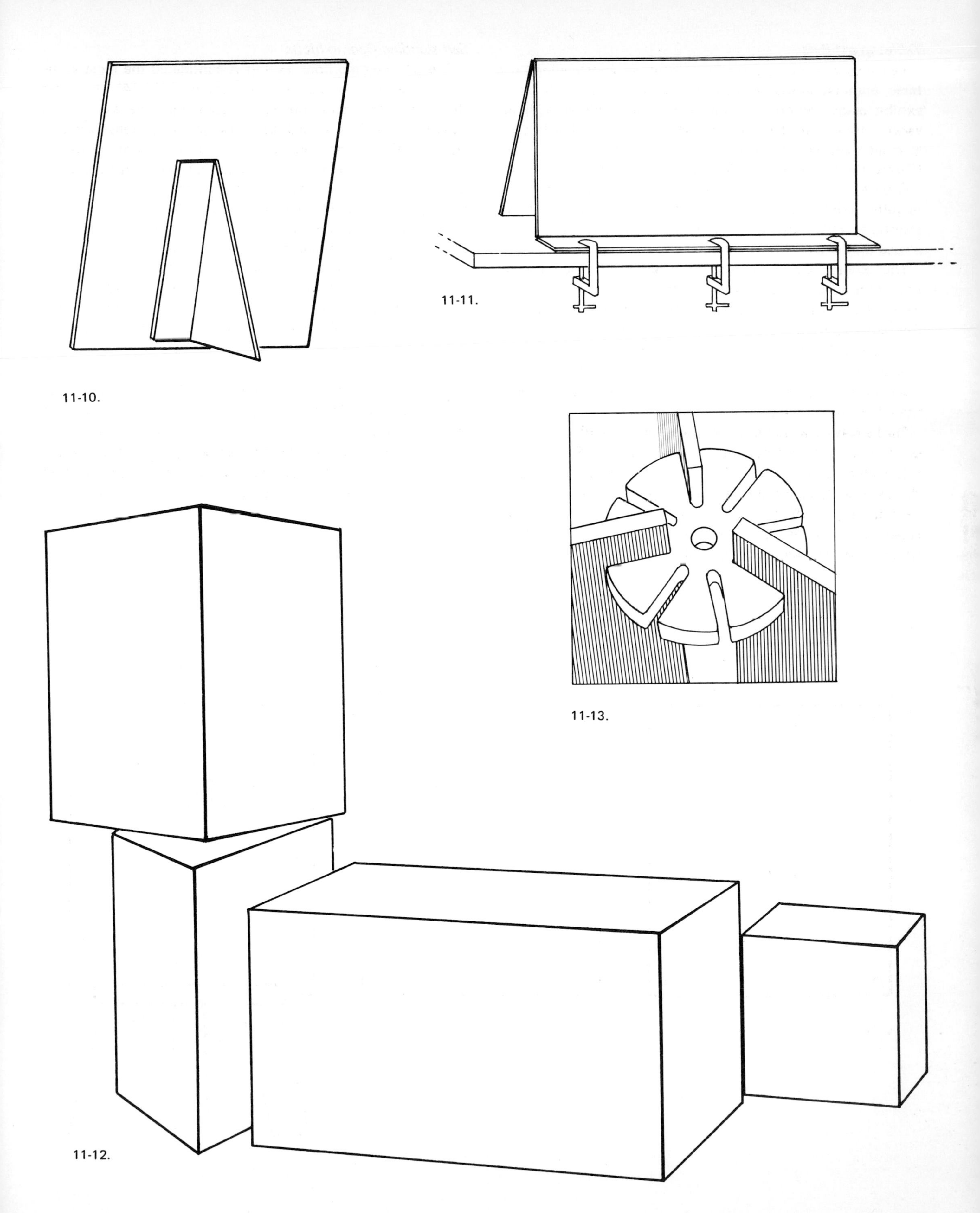

11-10.

11-11.

11-12.

11-13.

Commercial systems

There are display kits available that are self-standing, modular units that can be adjusted to various shapes and sizes. They can be taken apart or the legs retracted to fit into specially made carrying cases. Most include panels to which you add your copy and pictures.

One of the simplest is the exhibit kit made by William Hayett, Inc. It consists of painted foam-core panels measuring 24″ × 24″, clear-plastic Panelocks that connect the panels, and a carrying case. The Panelocks can be purchased separately and used with 3/16″ foam-core board cut to your own design (11-16).

The Exposystems Multiscreen kit consists of three-part panels measuring 2′ or 3′ in width and up to 7′ in height, which are covered with burlap or "touch and go" loop fabric. Metal connectors and framing are also supplied. All the pieces are separable and fit into an Expocase® (11-17).

There are many systems of components that can be arranged into almost any configuration. They consist of metal or plastic connectors that accept metal tubes cut to your specifications. System Abstracta is one of the brand names. They produce very sturdy and attractive structures on which panels and shelves of foam core, cardboard, plastic, or pressed board can be attached. They are designed primarily for exhibits of long duration and are rather expensive and difficult to assemble or disassemble (11-18).

11-14.

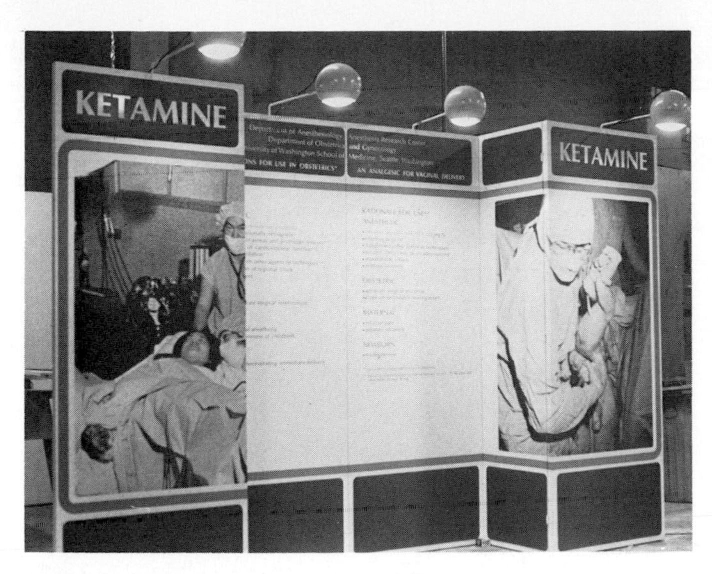

11-15.

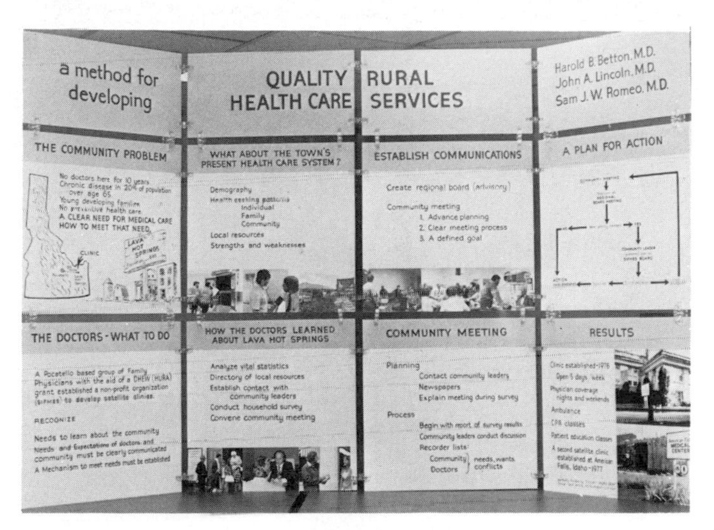

11-16. Carol Jerome.

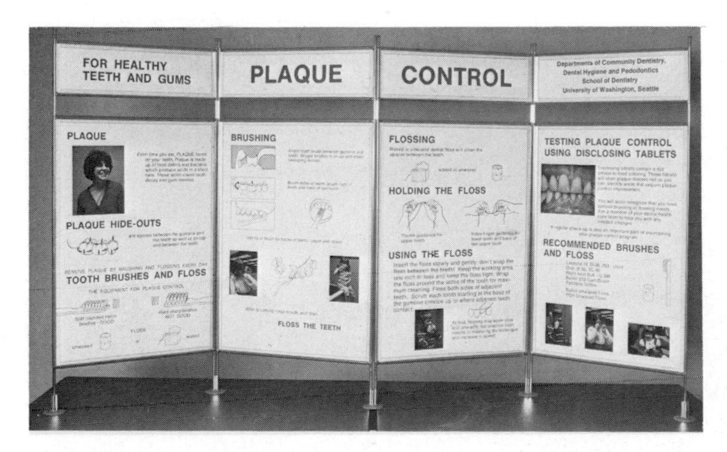

11-17. Virginia Brooks.

11-18.

Printed material for distribution

Much exhibit information would have a greater and longer-lasting impact if printed material were available at the exhibit. As the exhibit is designed mainly to stimulate further interest in a subject, interested viewers can use the handouts for more leisurely study and as educational or reference material.

Printed handouts may take many forms. The simplest is a single mimeographed sheet, and the most elegant is a four-color coated-paper brochure. In between are folded leaflets, pamphlets, and saddle-stitched booklets reproduced by duplicating machines on paper color-coordinated with the exhibit or printed in two colors. They may include detailed drawings and descriptions, instructions, references, and the names and affiliations of the scientists involved.

More time, money, and effort may be justified on the design, paper, and printing of the handouts if they are planned for multiple use and distributed at other times and to other people than exhibit viewers. They can be used as part of a class syllabus or continuing-education program.

Packaging and transportation

Packaging and transportation are part of the total scope of exhibit design and vary with the requirements of the individual project. An exhibit that is to be transported as baggage needs special protection; an exhibit that the client carries only a few miles also needs to be consolidated and protected from damage during transport.

The simplest poster-session exhibit, constructed of heavy sheets of paper that are to be tacked up, may be laid flat in the client's suitcase or packaged in a custom-made cardboard folder or box. Tying cloth tape around it to form a handle makes it easier to carry. All flat-hung exhibits can be packaged this way. Be sure to use stiff cardboard, not wrapping paper. The cardboard helps keep the corners from bending. The corrugated boxes in which poster board is delivered can be cut down to make a custom box to fit the exhibit.

Many exhibit kits come with specially designed transport cases. They are ideal, as they are usually light enough to be hand-carried by the client and checked with his baggage.

Box manufacturing companies make boxes to order. They should be made of heavy corrugated cardboard in two pieces—a top and a bottom that nests in it. You can purchase two heavy woven straps to hold the box together and a suitcase handle that attaches to the straps so that the box can be carried and checked like a suitcase (11-19).

Heavy or large exhibits must be crated professionally and are usually trucked to their destination. A period of 1 to 2 weeks is usually required.

The strength and protection capacity of the packaging depends on whether the package is going in the client's car or by commercial airline or truck. Commerical baggage handling must be assumed to be rough, and packing should be made accordingly. The panels and other components must be arranged so that they do not move around in the package, marring other components. They should be packed snugly and separated from each other at least by a sheet of paper. If the exhibit has many panels, it should be arranged so that the first pieces taken out of the case are those that are put up first. The exhibit should be identified clearly both inside and outside with the client's name, address, and destination.

Assembling, dismantling, and storing

Never assume that the exhibit that you have designed and constructed will be assembled in the way that you have planned until you have put it together yourself. Any mistakes or difficulties should be recognized and corrected in the studio instead of on the exhibit floor, where they may be difficult or impossible to rectify.

It is up to the artist to demonstrate to the client how to put the exhibit up, take it down, and pack it into its carrying case. The artist should include with the exhibit a simple sketch of the assembly and a numbered list of instructions for putting it up.

It is generally the responsibility of the client to find storage for the exhibit. After an exhibit shows signs of wear, it should be put to rest, as a shop-worn exhibit gives a negative impression and detracts from the impact. The information on an exhibit is generally of timely significance anyway and loses its impact quickly. If the exhibit is no longer usable, some of the components may be recycled for another exhibit. Exhibit kits with multiuse panels should be reused, spreading initial investment costs, and not stored with out-of-date information on them.

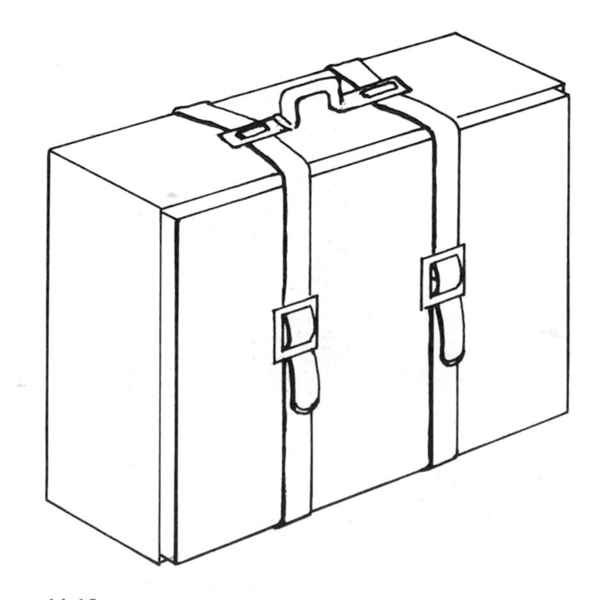

11-19.

DESIGN

After all the preliminary planning has been done, the actual design is worked out. This includes the logical placement and scaling of the copy and pictures and any design elements needed to tie them together or to control eye-flow direction.

Attention

Even a low-key, factual scientific exhibit needs some huckstering or commercialism to attract attention. It is difficult to attract viewers with an uninterrupted area filled with tables, charts, and copy that are similar in size and impact and that have no color, no matter how fascinating the subject. Such an exhibit will not accomplish its purpose, to communicate information and to stimulate viewer interest.

Color

Valid uses of color in an exhibit include pointing out important or related aspects of a subject, guiding eye flow, and setting a mood. Blue is wet, serene; red evokes pain, demands attention; yellow is cheerful, optimistic; clear, contrasting colors are active, stimulating. A black-and-white exhibit, however, can be very dynamic. A very small amount of color, used judiciously, dramatically focuses attention to important elements.

Pictures

Pictures are effective attention-getters. Well-executed drawings or photographs, in color or in black and white, can be either informative or mood-setting. Scale is important. A photograph blown up to full panel size, either with or without copy, can be very dynamic (11-20). Drawings should be rendered with the viewing distance in mind.

Three dimensions

Another attention-getter is a three-dimensional construction. The actual shape of the exhibit can be stimulating. The same quality can be conveyed in a flat exhibit by using three-dimensional models. Photographs, either rectangular or silhouette, can be mounted on foam-core board, as can bars of charts and parts of pie charts (11-21).

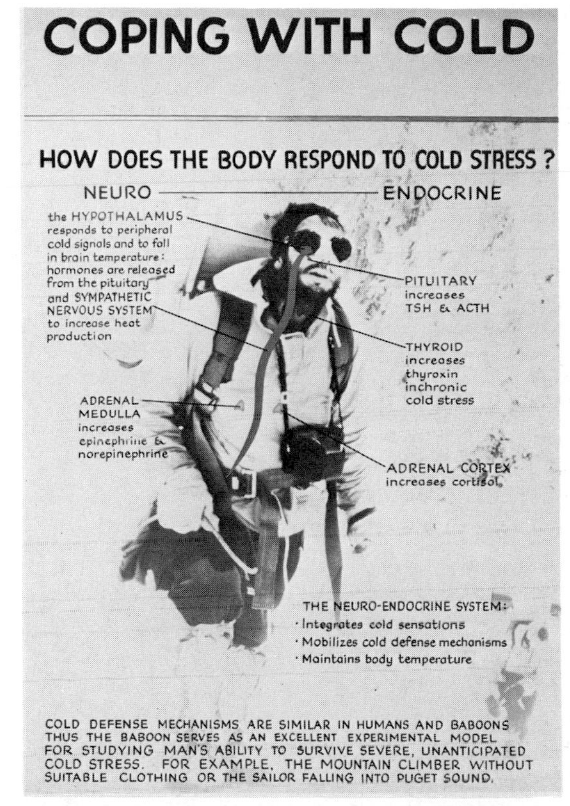

11-20.

11-21. David Morris, designer.

Lighting

Backlighting transparencies or radiographs can add extra interest. Spotlighting the front of the exhibit may also be effective.

Slides and video

Slides in a rear-projection screen or a continuous television tape with or without sound stimulate interest and can be used if they add to the informative qualities of the exhibit. Sound, lights, and movement must be used with discretion, however, as they may only confuse or annoy the viewer instead of attracting him.

Eye flow

After being attracted the viewer will probably take a quick look over the exhibit to see whether it warrants further inspection; if he is interested in the subject, he will examine it in more detail, conventionally starting at the upper-left side. The design of the exhibit, through the use of color and the directional thrust of the copy and pictures, should guide his eye. The eye-flow direction can run down each panel vertically in sequence or horizontally across the area, depending on the size of the exhibit. The viewer should move in only one direction, from left to right, or not at all (11-22).

Lettering

Lettering must be done in a legible size, style, and arrangement.

Size

The size of the lettering is determined by the distance from which it is read. The main title of the exhibit, usually above head level, is meant to be viewed at a greater distance and the lettering should be in capitals 2″ to 4″ (5 to 10 cm) high. Heads and subheads can vary from 1/3″ to 1 1/2″ (9 to 40 mm). Body copy should rarely be smaller than 1/3″ (1 cm). Using only a few type sizes contributes to a coordinated appearance (11-23).

Arrangement

All the type should be arranged in a logical sequence for readability. When planning the body copy, avoid hyphens and widows (a single word on the last line of a paragraph). It is not necessary to justify the type—to have right and left margins flush (11-24).

Style and methods

You can never go wrong with Helvetica medium, but, if you want to be more innovative, many typefaces may be appropriate. Think of the word that best describes the subject of the exhibit and match it with a typeface that has the same personality (11-25).

Legibility is the most important attribute of any exhibit typeface. Decorative or avant-garde styles may be an ego trip for the designer but a puzzle for the viewer. Two typefaces should be considered the maximum; one typeface may be preferable.

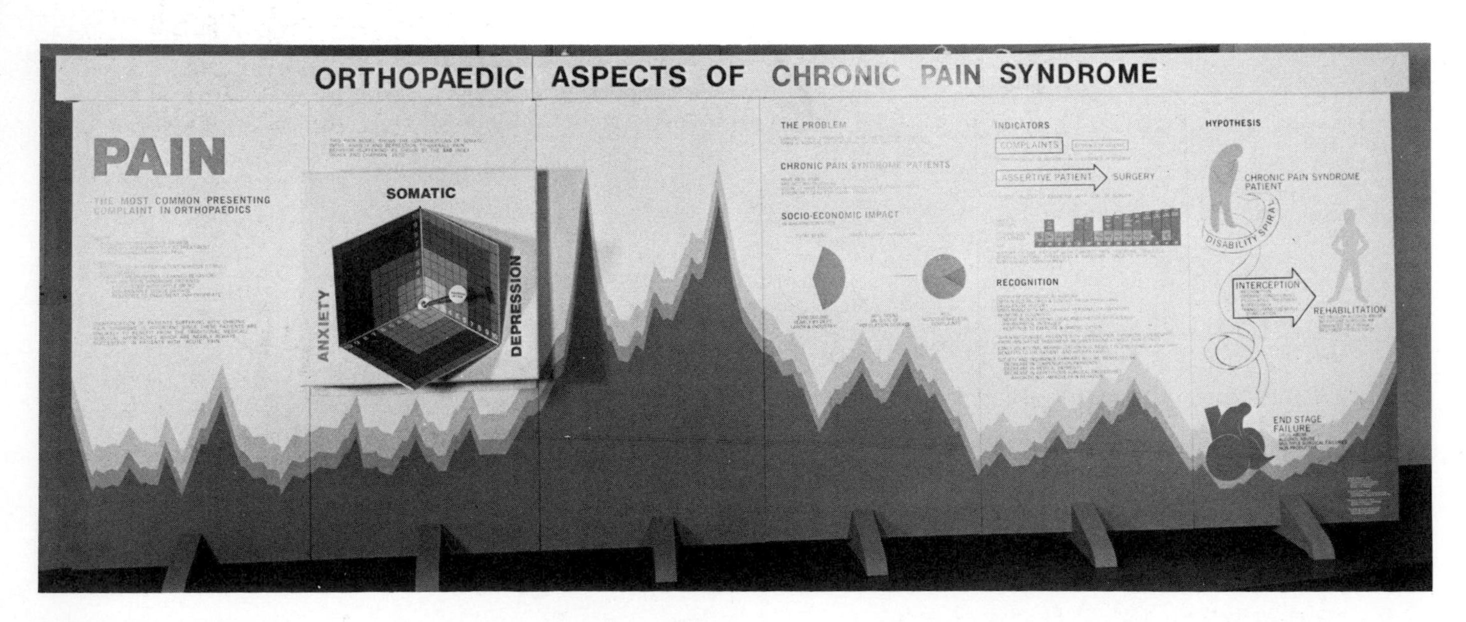

11-22.

PREPARATION OF BUFFY COATS FOR ELECTRONMICROSCOPY

PATRICIA MANNING W ELLIS GIDDENS

REGIONAL PRIMATE RESEARCH CENTER
UNIVERSITY OF WASHINGTON
SEATTLE WASHINGTON 98195

The buffy coat technique
allows the concentration and preparation
of large numbers of circulating leukocytes
for electronmicroscopy.

We use this technique to examine lymphocytes
from owl monkeys (Actus trivirgatus)
infected with Herpesvirus saimiri
for the presence of virus particles.

Heparinized whole blood is collected
from leukemic and control monkeys,
and processed in the following manner.

References
Anderson, D.A.J., Ultrastructure Research 13, 263-268 (1965)
Giddens, W.E., International Academy of Pathology, 32,492 (1975)
*Microfuge tubes, Beckman Instruments Inc.

1
Chill blood in ice.
Place in 0.6 ml plastic centrifuge tubes*.
Centrifuge at 800 x g for 10 min at 5° C.

2
Remove plasma carefully.
Place fixative (1% glutaraldehyde in 0.1 m sodium cacodylate buffer) in tube over buffy coat.
Fixation time is 30 min at 5°C.

3
Remove fixative.
Carefully cut tube above and below fixed buffy coat.
Gently remove plastic ring.
Hold in 0.2 m sodium cacodylate buffer
with 7% sucrose at 5° for 1-7 days.

4
Post-fix disc 1 hour
in 2% osmium tetroxide in s-collodine buffer.
Dehydrate in graded ethanol solutions
and propylene oxide.
While disc is in 80% ethanol,
trim into pie-shaped pieces
approximately 0.5 cm in size.
Complete dehydration.
Embed in coffins with Epon 812.

5
For general observation,
stain 1-μm sections with Toluidine Blue.
For electronmicroscopy,
prepare silver-refractive thin sections
on uncoated copper grids.
Stain with saturated aqueous uranyl acetate
and lead citrate.

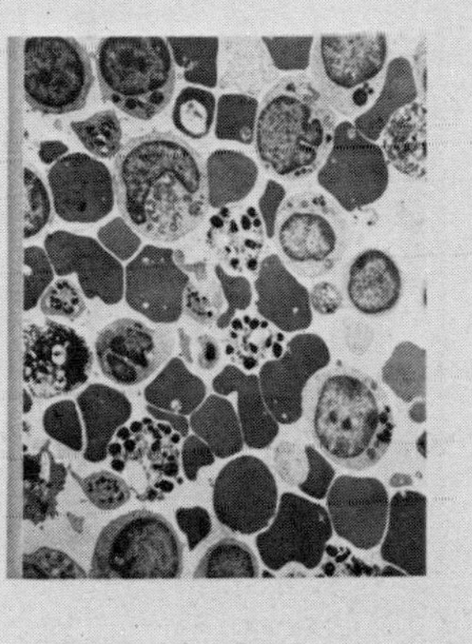

11-23.

It is usually better
to arrange each line
flush left and ragged right
so that they read
in a natural phrasing rhythm.

Justifying the type may lead to awkward phrasing, whose meaning is not quickly recognizable.

11-24.

Mycological
Ornithological
Contemporary
Traditional
ENVIRONMENTAL

11-25.

There are many ways to letter the exhibit panel. They vary greatly in cost, attractiveness, and needed time and tools and must be chosen on a cost-effective basis.

Silkscreen

Silkscreen is the handsomest method of producing lettering (11-26). It can be done directly on large or odd-size panels, in one or several colors, and on any material—plastic, metal, wood, glass, foam-core board, or cardboard. The screens can be used many times, making this an inexpensive method of producing multiples such as posters. It is an expensive but polished method for a one-copy exhibit.

Have the copy typeset or lay it out with dry-transfer lettering. It can be accompanied by any desired design effects (borders, lines, arrows), done in black or opaque red. This lettering layout does not have to be in the same size as that of the lettering planned for the exhibit—it can be larger or smaller. Type that is to be enlarged must be examined very carefully for any irregularities, as the irregularities will also be enlarged and become more obvious. Clean, crisp phototype can be greatly enlarged.

The type is mounted on white mounting board, and instructions written for needed reductions or enlargements to match the sizes planned for the exhibit panels. If more than one color is used, each color must be prepared and mounted separately. From the black-and-white paper copy exhibit-size film positives are made. A film positive is a duplicate of your black-and-white copy on film. Film size ordinarily reaches a maximum of 20″ × 24″ (50 × 60 cm).

The film positives are arranged and spliced together so that they fit the panel design. Use heavy tracing paper for the blank areas and attach the film together with transparent tape so that it is the same size and arrangement as your panel layouts. This should be done in cooperation with your silkscreener. The limit to the screen size varies with the individual screener. Many sign companies do silkscreening, but quality varies, with some preferring large-quantity jobs and others small specialty jobs.

From the film positives the silkscreener makes a stencil photomechanically on fine cloth or stainless-steel mesh, which is mounted in a frame. Heavy paint is pressed through the fine mesh onto the exhibit panel with a squeegee. This method can be recognized by the heavy layer of paint and the perceptible mesh pattern (11-27).

Handlettering

Handlettering can be divided into two distinct qualities, professional and amateur (11-28). Unless you are trained and practice constantly, you are an amateur. If you wish to hire a professional sign painter to do your lettering, lay it out on a 1:1 scale (actual size) in the desired arrangement. This is the job of the designer. It can be done in pencil or with felt-tip marking pens. The sign painter follows your instructions as to color and organization, reproducing your lettering in either oil-base or water-base paint. Unless an exhibit is to be of a permanent or outdoor variety, the lettering is done in water-base or show-card paint.

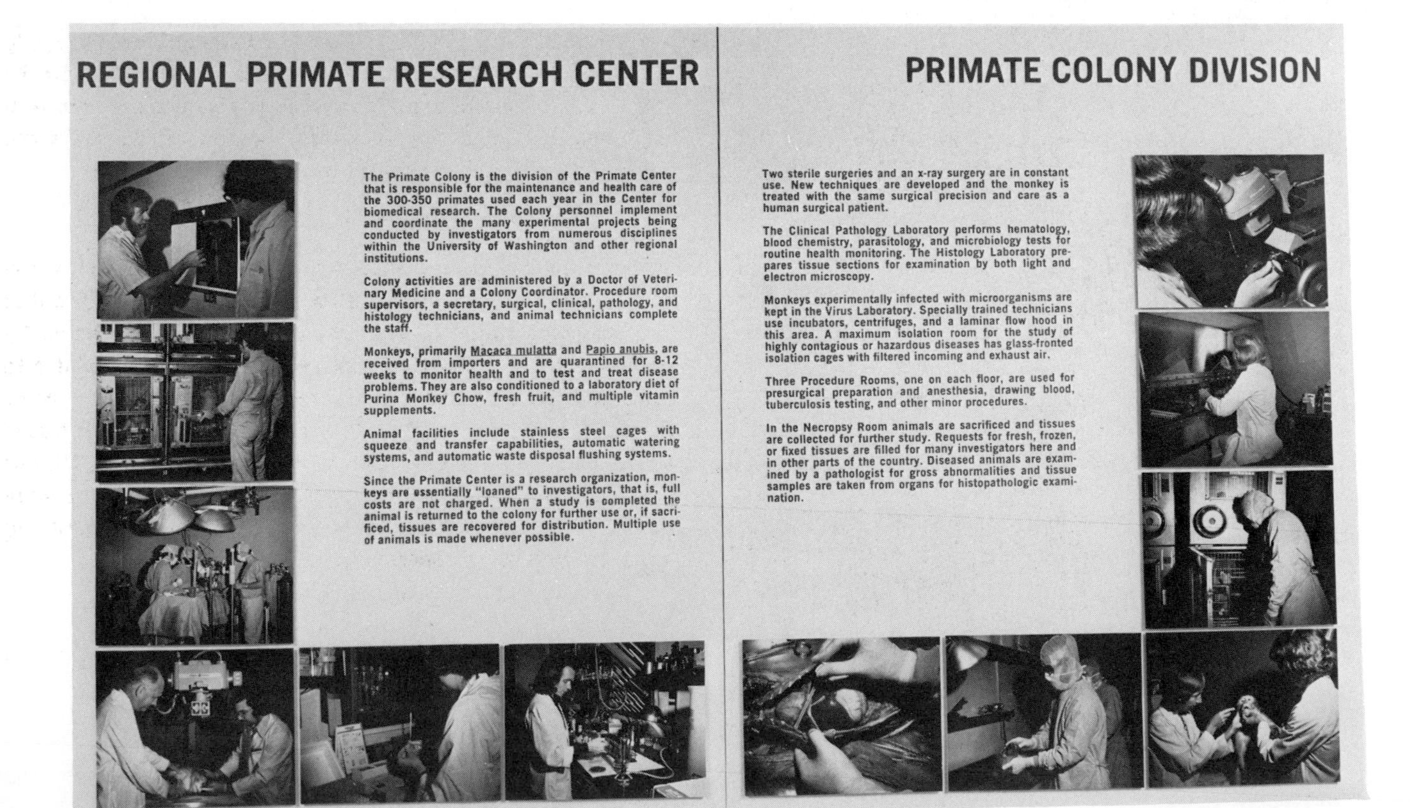

11-26.

11-27.

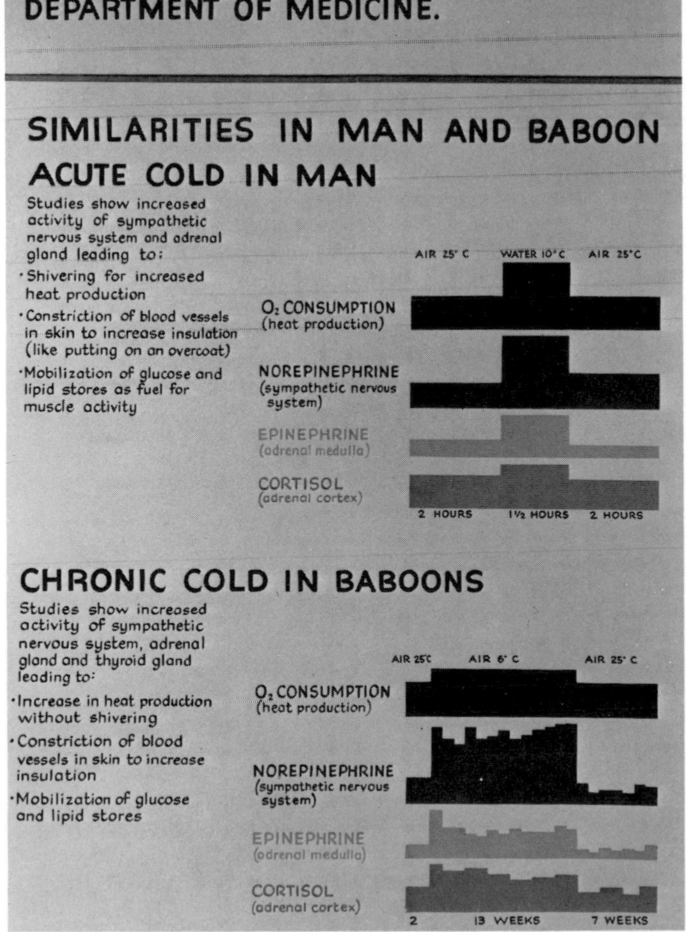

11-28.

If you plan to do your own handlettering, you can use either lettering brushes and poster paint or felt-tip marking pens in different widths. Make light guidelines and follow them carefully. Practice with the same colors and type of tools and board as for the final product.

Dry-transfer lettering

Dry-transfer lettering, while tedious and expensive to apply in any quantity, gives a crisp, professional look. It is fragile and mars easily, but spraying with mat fixative after burnishing helps to protect it. It comes in black, white, and some colors and in a great variety of styles and sizes.

Cutout lettering

If you are doing very limited titles, cutout lettering is versatile and attractive. Assemble all the letters in the typeface that you are going to use and have them enlarged photographically on heavy paper. Cut them out and use them as templates to trace each letter. They can be traced onto paper of any color or texture to coordinate with your exhibit. Glue the traced letters onto the exhibit panels with rubber cement or Double Nothing. If a few dramatic three-dimensional letters or numbers are desired, they can be cut out of foam-core board and attached with foam-adhesive tape.

Mechanical lettering

Using a LeRoy lettering set, you can letter directly on the exhibit panels with black or colored ink. It is a neat if not elegant method. Corrections are not easily made on color backgrounds (11-29).

Photocopied lettering

Any black lettering can be enlarged photographically onto white mat-finish paper and mounted on the exhibit panels. Typed copy, LeRoy lettering, phototypeset, and dry-transfer lettering can all be used this way. The watchwords are "black," "crisp," and "clean." Any imperfections in the type will be enlarged, and gray type may fall out completely. The white photographic paper becomes part of the design. Many white pieces of copy may give a spotty appearance if they are not coordinated with the overall plan of the exhibit (11-30).

Die-cut lettering

Die-cut paper letters can be punched out on a Leteron machine or purchased. They come in several colors, sizes, and styles. Plaster, cork, and wood lettering is also available from a display house or catalog.

PANEL MATERIALS

Commonly used panel materials have different characteristics and must be handled and assembled within their capabilities and limitations.

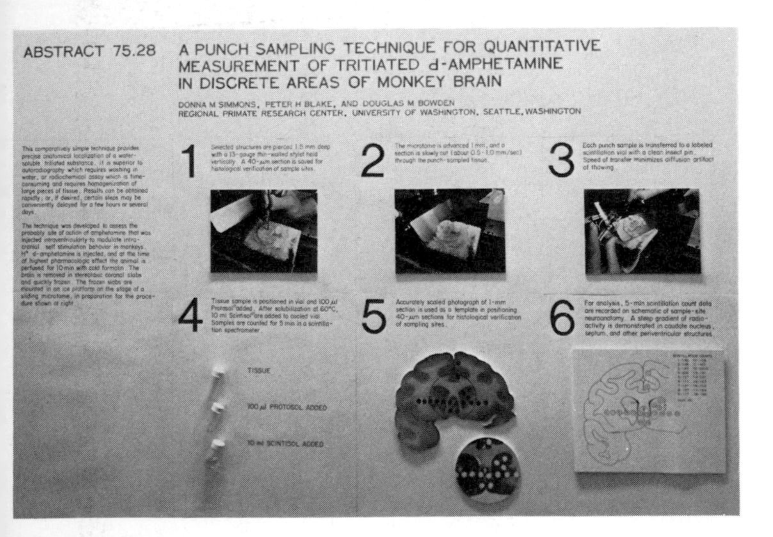

11-29. Robyn Tarbet, designer.

Foam-core board

One of the most favored panel materials is 3/16″ foam-core board, because it combines light weight with rigidity. It is a sandwich of Styrofoam between smooth white paper facings. It also comes in thicker and thinner grades. Sizes run from 20″ X 30″ (50 X 75 cm) to 48″ X 96″ (125 X 240 cm). By buying a carton of the large size you get a generous price break. The slick surface of the board should be roller- or spray-painted on both sides with a washable paint before you letter or mount on it.

Foam-core board is easily cut with a very sharp-pointed scalpel or mat knife held almost vertically and guided with a metal straightedge. If the cutting edge is dull, it may tear the facing paper or compress the foam. Try to cut cleanly through with one stroke. The board can be scored carefully through one facing paper and bent in the opposite direction. It can be taped together with Mystic tape, a very strong, permanent cloth tape that comes in white and several colors and in several widths. Mystic tape can be run along the cut edges to protect and seal them. If Panelocks are used to join the panels, slightly compress the edge of the 3/16″ (5 mm) foam-core board with your fingers to allow for the smaller opening of the Panelock without tearing the facing paper. The Panelock does leave an impression in the board, so use the same spot to connect the panels each time. Care must be taken when working with foam-core board, as the corners can easily be crushed and sharp objects can penetrate the facing paper and damage the foam. It is not repairable!

Mat board

Textured, mat-finish Crescent or Bainbridge cardboard in a double thickness—approximately 1/8″ (3 mm)—measures 32″ X 40″ (80 X 100 cm). It comes in many beautiful colors on one side only and is ideal for flat-hung exhibits and posters.

Thin cardboard

For poster sessions in which the exhibit is divided into small increments and tacked to a wall use thin cardboard or heavy 2- or 3-ply paper. Strathmore plate or kid-finish is a good choice, as is Husky Blank. These boards come in white only. Railroad board or index card may be used, and it comes in many colors.

Pressed board

Pressed board (Masonite) is very strong and heavy. It is best used in a permanent exhibit where portability is not a factor.

Plastic

Plastic panels (Plexiglas) 1/16″ or 1/8″ (1.5 or 3 mm) thick can be used effectively to exhibit material hung on the wall, in backlighted radiological view boxes, or in a commercial module. Plexiglas measures 40″ X 60″ (80 X 150 cm) and can be used either full- or partial-size, depending on portability requirements. All the graphic material is taped in place, silk-screened, or painted on the back of the Plexiglas, creating a protective and slick, attractive front surface.

Plexiglas comes with a protective paper cemented to the back. It can be cut with a blade and peeled away in sections. Three coats of opaque rubber-base or acrylic paint applied to this peeled surface by spray or roller will produce a perfectly smooth appearance on the front side. Silkscreening may be applied on the reverse side. Informative material (drawings, graphs, written copy) can be taped to the windows produced by the painting process.

If the exhibit is to be backlighted, radiographs, color transparencies, or film positives or negatives can be used effectively. The construction of transilluminated panels must be integrated into the design format. The exhibit panels are constructed as boxes 6″ to 8″ (15 to 20 cm) deep, with the fluorescent tubes on the back wall for even lighting with no "hot spots." Easy rear access for installation and replacement of the tubes is necessary. Special care is required in packing and shipping.

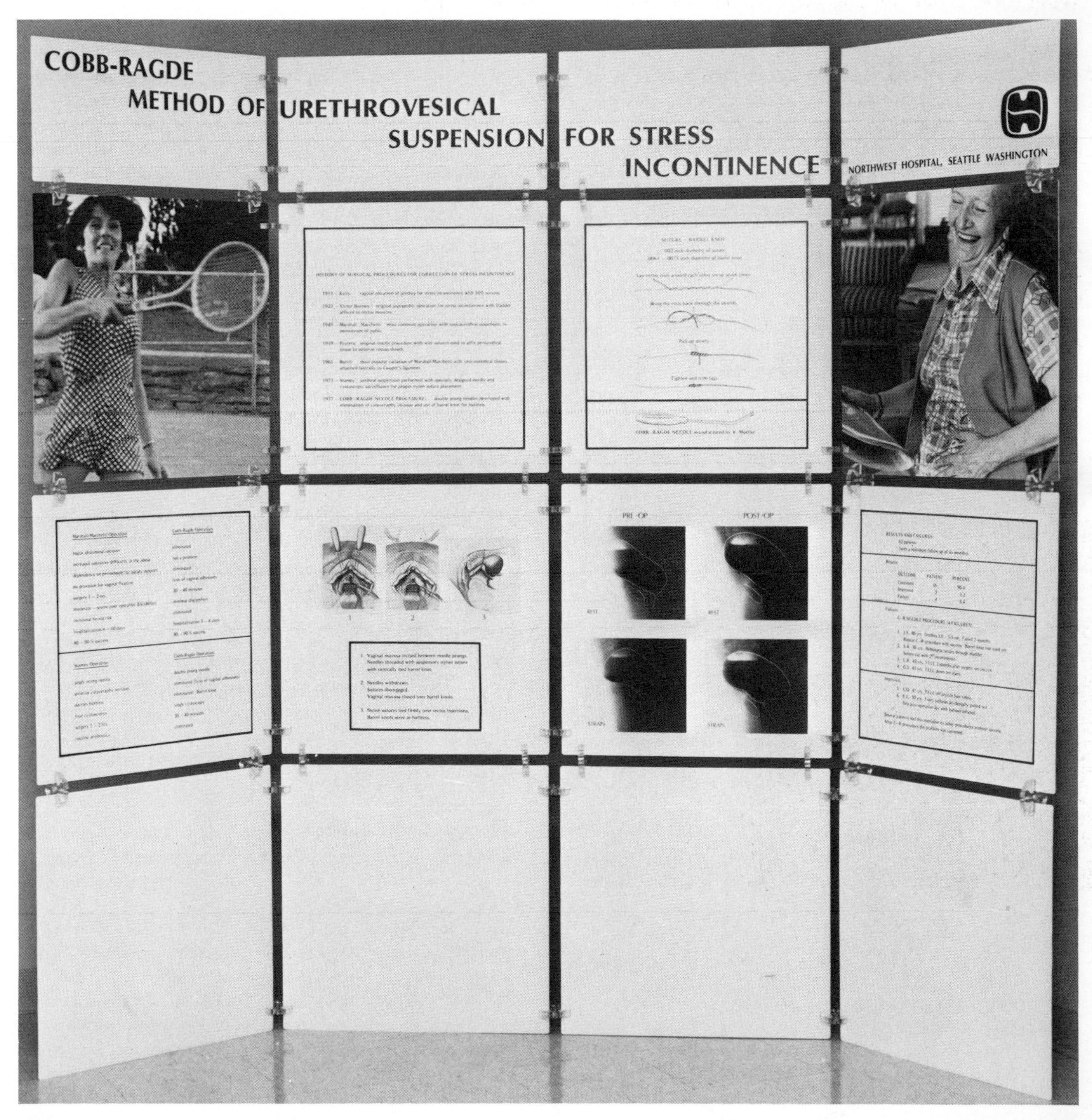

11-30.

MOUNTING

All informative material that is not painted directly on the exhibit panels—photographs, drawings, diagrams, and written copy—must be mounted securely in one of three ways: flat, elevated, or window.

Flat mounting

In flat mounting the graphic material is cemented directly to the panel, maintaining a level surface (11-31). This is the easiest type of panel to pack and transport.

Elevated mounting

In elevated mounting the graphic material is mounted first on a piece of foam-core board and then on the exhibit panel (11-32). The three-dimensional effect and the cast shadows caused by the elevated layer add design interest and direct the viewer's attention. The three-dimensional look can be heightened by using several layers of foam-core board or by cementing additional pieces of board on the back before affixing it to the panel with foam-adhesive tape.

Window mounting

Windows can be cut in the exhibit panel and the graphic material taped on the back with Mystic or masking tape (11-33). This creates a nice three-dimensional effect, eliminates the need to crop the graphics, and pulls all the graphics together, especially if they are diversified in size, style, and subject. The graphics can be protected with an acetate flap inside the window. The more the windows and the closer together they are, the weaker and the more subject to damage the panel will be. This must be considered in the construction and hanging plans.

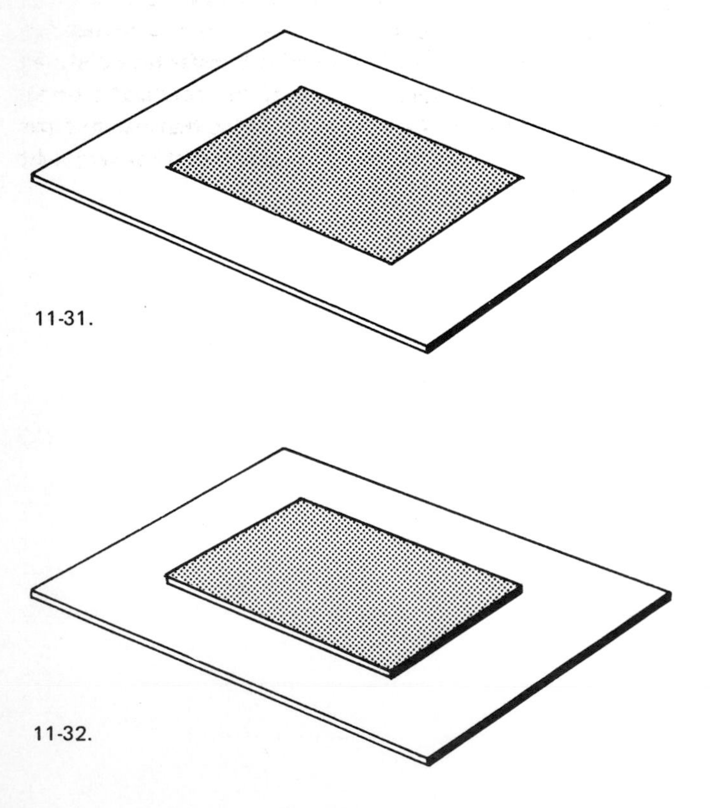

11-31.

11-32.

Mounting materials

The materials used for mounting vary in strength, longevity, and flatness.

Rubber cement

Rubber cement is easily available and produces a strong, smooth mount if used correctly. It may seep through and stain the mounted material after a number of years. Rubber cement thickens with exposure to air, so the container should be kept tightly closed. It can be thinned with rubber-cement thinner. The bond can be dissolved by gently lifting a corner of the paper and applying rubber-cement thinner with a brush or spouted can (Valvespout® dispenser). By slowly lifting and separating the paper from the mount while applying the thinner you can remove the mounted material without destroying either it or the mount. Spray adhesive can be used as a satisfactory alternative to rubber cement. Following is the procedure for rubber-cement mounting.

1. Trim the photograph or other paper material to be mounted.
2. Lightly outline the mounting position.
3. Apply a thin coat of rubber cement to the entire outlined area on the mount.
4. Apply a thin coat of rubber cement over the entire back of the photo.
5. Leave both coats of rubber cement until they are dry to the touch.
6. Line up the top of the photo with the top edge of the outline on the mount and smooth the top edge into position.
7. Gradually allow the photo to roll into position from the top down, smoothing and pressing with your hands as you go.
8. Press and rub the entire surface by hand to secure the bond.
9. Any excess rubber cement can be removed by rubbing with your fingers or a rubber-cement pickup, usually without staining the mount or the photo.

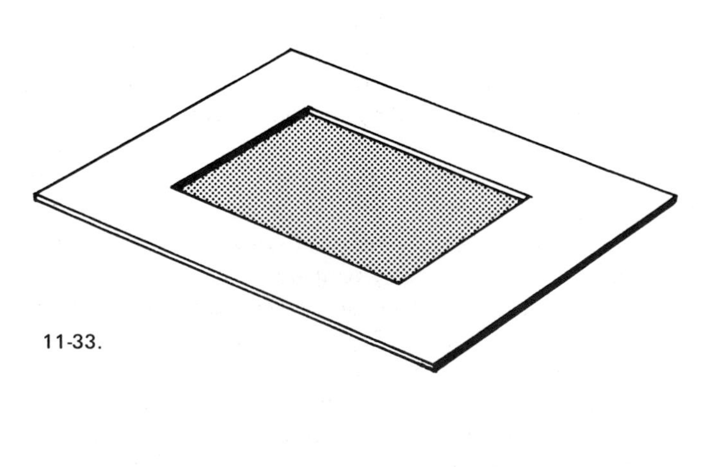

11-33.

Double Nothing

Double Nothing is the common name for various adhesive papers consisting of a layer of pressure-sensitive adhesive sandwiched between two layers of protective release paper. It is expensive but produces a beautifully smooth and strong bond. The bond can be dissolved with rubber-cement thinner. Following is the procedure for mounting with Double Nothing.

1. Cut a piece of Double Nothing slightly larger than the photograph or other paper material to be mounted.
2. Lift one side of the release layer from the adhesive almost completely and lay it back in place except for the top edge, which should be turned back in a 1″ (3/8 cm) cuff.
3. Line up the top edge of the photo with the top edge of the adhesive surface above the cuff and smooth them together.
4. Slowly peel the release layer away, at the same time gradually smoothing the photo into position. Avoid trapping air bubbles between the adhesive and the photo.
5. Trim the photo and the Double Nothing together.
6. Mark a light outline on the mount where the photo is to be mounted.
7. Lift the release layer from the back of the photo almost all the way and lay it back in place except for the top edge, which should be turned back in a 1″ (3/8 cm) cuff.
8. Line up the top edge of the photo above the cuff with the top edge of the outline marked on the mount and gently smooth it in place.
9. Gradually peel the release layer back, at the same time smoothing the photo in place and avoiding air bubbles.
10. Air bubbles that are trapped between the photo and the mount may be pressed to the edge with your fingers or released by pricking the surface of the photo with a sharp instrument.

Dry mounting

Dry mounting produces a very flat, permanent bond that does not discolor or stain the mounted material even with age. In dry mounting wax bonding tissue is sandwiched between photographic or other paper to be mounted and the mounting board. The layers are fused by heat and pressure applied with a dry mounting press or, for small jobs, a hand iron. The temperature and time required depend on the thickness, the heat-conducting characteristics, and the emulsion surface of the paper to be mounted. Some photographic emulsions melt under ordinary dry-mounting temperatures. Dry-transfer lettering melts and disintegrates under any heat. Wax-back adhesives release their bond. Test a sample before diving into your only copy. If you are mounting a paper larger than the area of the press, start in the middle and work out to the edges to avoid air bubbles. The dry-mount bond can be dissolved with alcohol, dry-mount solvent, or rubber-cement thinner. Following is the procedure for dry mounting.

1. Place the photograph or other paper material to be mounted and the mounting board in the heated dry-mount press to flatten and dry. If this is not done, they may dry differently, causing a permanent curl or preventing a strong bond.
2. Place the photo face down with a piece of dry-mounting tissue over it.
3. Make a small x in the center of the tissue with a heated tacking iron, fusing the photo and the tissue together at one point.
4. Trim the photo and the tissue together to the same size.
5. Mark a light outline on the mounting board where the photo is to be mounted.
6. Place the photo and tissue, tissue down, inside this outline.
7. Holding the photo securely in place with one hand, lift each corner of the photo and tack the corners of the tissue to the mounting board with the tacking iron.
8. Place the photo and mount face up in the heated dry-mount press, protected by a clean cover paper. Leave it under pressure for the time and temperature previously selected.
9. Remove from the press and place under a flat weight until cool.

Tapes

Included among the adhesive tapes that are used behind the scenes are: Mystic tape, which is very strong and dependable; masking tape, strong but dries with age; and transparent tapes, least strong and dependable.

Double-coated tapes can sometimes be used in place of other bonding agents. Double-coated foam-adhesive tape is 1/16″ thick and is durable and strong. It can also be purchased in packages of 1″ squares. Scotch adhesive-transfer tape designed for use with the ATG dispenser is easy to apply and satisfactory for light mounting jobs. Double-coated transparent tapes and masking tapes are not recommended except for very light and temporary mounting jobs.

CHAPTER 12.

Career Guide

Preparation for a career in the scientific-illustration field involves more than merely acquiring technical proficiency. Building a reputation and a clientele is a continuing process.

EDUCATIONAL BACKGROUND

In addition to a thorough grounding in illustration and reproduction techniques the education and training necessary to a scientific-illustration career include at the minimum a bachelor's degree in one of the scientific fields. It is certainly advantageous to pursue a postgraduate degree in the field of your choice. While a competent illustrator has the skills to illustrate what he sees in any medium, he may not know what he sees unless he has a knowledge of the subject. He must see his subject in the broader context of its relation to conspecifics and to other specimens. He must see shapes as they relate to structure and function. It is for this reason that most scientific illustrators specialize in a subgroup in the fields of zoology, botany, geology, forestry, anthropology, medicine, or dentistry.

In the medical-dental field there are six schools that offer accredited masters-degree programs. Information may be obtained from the Association of Medical Illustration, 6650 Northwest Highway, Chicago, Illinois 60631. There are a number of nondegree training programs in scientific illustration. Information may be obtained from the Guild of Natural Science Illustrators, P.O. Box, 652, Ben Franklin Station, Washington, D.C. 20044.

THE PORTFOLIO

Building a portfolio is one of the most important parts of your job preparation. Organizing current examples of your artwork and writing your curriculum vitae are a continuing process that grows and develops as your skills mature and your interests and specialties change. Your portfolio should be a carefully planned reflection of your artistic abilities.

Contents

Examples only of what you consider your best work should be included. Quality, not quantity, is important. It is best to include drawings in the final form for which they were designed (print or slide), as the reproduction quality, rather than the original artwork, is significant. If original drawings have not been reproduced, have them reduced to the size in which they would appear in print form. Use a quality photocopying or photographic process, not a machine copy.

If you are proficient in several techniques, include one or two good examples of each. Vary the examples in your portfolio as to subject and technique according to the prospective client. Include more examples of subjects similar to those in his field, illustrated in a technique that would be suitable for his publication or projection needs.

Presentation

While there are many elegant styles of presentation books and folders, the most important requirement is that they be *neat*, *clean*, and *precise*. Every mat should be cut precisely, crisply, and at right angles. All corners must be sharp and uncrushed. Everything white must be virginal; every mounting must be flat and firm. Every reproduction must be of excellent quality. There should never be a reason to apologize for the physical condition of your portfolio. A neat, clean presentation makes average drawings look good and good drawings look excellent.

While you can custom-design and -construct your presentation portfolio, there are several styles and many sizes that can be purchased or ordered through art stores. Zippered presentation books with wire bindings come with acetate pockets to protect your illustrations. Folders come in heavy vinyl "leather" with zippered closings or in sturdy cardboard with ribbon ties. Some of these are available in sizes up to 30″ X 40″ (12-1).

Illustrations should not be carried loosely in the portfolio or in the acetate pockets. They should be mounted on pieces of black mat-finish illustration board all the same size. Black is used because it is completely neutral and does not modify the illustration. The mat finish is used to eliminate glare. Same-size mounting boards are used to present a unified appearance. One or several related illustrations can be mounted on the same board (12-2). The illustrations can be mounted flat on the surface of the boards, or you can cut windows in the board and mount them from the back. In this case the back should be covered for a neat appearance.

When you are applying for an illustration job by mail, it is awkward to mail a portfolio, and it is liable to be mislaid in the mail or not to be returned properly or promptly by your prospective client. In this case you should have some good reproductions made of your illustrations. Photocopies, photographs, and offset-duplicator copies can be of good quality. Be critical of the reproductions. Remember that you will be judged solely on the visual impact that they make. Organize the mailing piece carefully. It can be as simple as stapled 8 1/2"- X -11" pages or as sophisticated as a custom-designed brochure.

Curriculum vitae

The Latin term *curriculum vitae* translates as "course of one's life." Webster defines it as: "a short account of one's career and qualifications prepared typically by an applicant for a position." The curriculum vitae is always included with the portfolio. It is ordinarily on the first page of the folder, book, or stapled mailing piece.

While much of your personal history can be included, it is necessary only to include the part that pertains to your job qualifications: education, positions held, and honors received. You should also include major art projects that you have been involved in and the extent of that involvement.

This information should be typed in an easy-to-read form by an expert typist. It should be added to and revised to keep it current.

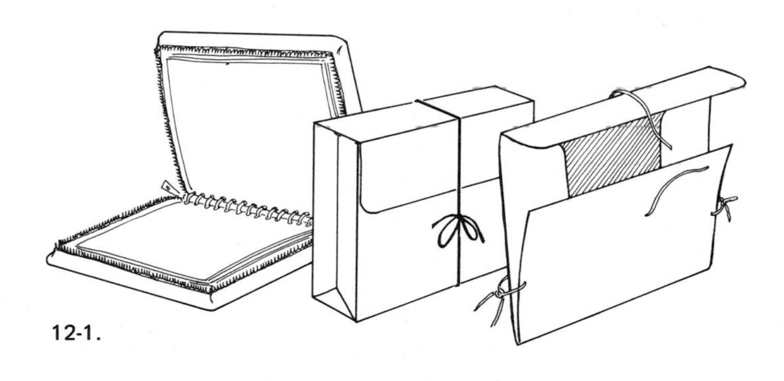

12-1.

12-2.

BUILDING A CLIENTELE

Building a reputation goes hand in hand with building a clientele, as satisfied clients are usually delighted to pass an artist's name along to their colleagues. The hardest assignment, of course, is to get the first few illustration jobs.

The first step is to find out where the job market for your particular specialty lies and to contact the prospective institutions, departments, and individual authors. There is a tremendous amount of writing going on at all times in all scientific areas, and, as Alice in Wonderland said, "What is the use of a book without pictures?"

Colleges and universities are natural places to look for illustration assignments, as are scientific foundations and museums of anthropology, natural history, and similar fields. The federal government is a large buyer of illustrations for its many publications; information may be obtained from a federal job-information center. The Civil Service Commission has procedures for grading and hiring applicants. There are also many independent publishers. Look at scientific books in your local bookstore and decide which ones might benefit from your services. Write to them and enclose your mailing portfolio.

After you have acquired a client, see that he is your unpaid public-relations agent. You can do this by delivering high-quality illustrations within the time period on which you have agreed. Never promise an illustration that you are incapable of producing either in time, subject, or technique.

BUSINESS PRACTICES

While *art* may seem completely unrelated to *business*, the artist, to be respected as a professional and to receive a fair return on his time and talents, must conduct his artistic endeavors on sound business principles. He must charge his client so that he makes a salary commensurate with his product in the current job market.

Estimating

It is best to estimate the cost of a project within a range rather than as a single amount. If the actual cost is at the high end of the range, the client will not be surprised, and if it is at the low end, he will be happy. It should be made clear to the client that this estimate does not include additions, changes, or extraordinary services. Nor does it include any changes made to the finished drawings after the client has approved the preliminary final sketches. The estimate for a job is computed on the basis of the hours needed to produce the artwork plus the expenses directly applicable to the project.

Hourly rate

The hours spent must be converted into dollar value. The five factors that determine this conversion are: time, talent, overhead, fringe benefits, and current salary range. You may find it easier to figure this on a yearly basis, proceeding from a proposed yearly salary, expected yearly overhead, and fringe benefits.

The time spent on a job should be recorded meticulously. Besides drawing-board time you should include time for conferences with the client, research, thinking, ordering material, consulting experts, travel, bookkeeping, and wrapping and mailing.

Each artist has a different level of artistic talent, production speed, background knowledge of the subject, and technique and style specialties. In other words, each artist's worth varies in the job market.

Overhead includes equipment and supplies, studio rent and utilities, and general research expenses. Equipment includes such items as drawing board, chair, light, paper cutter, lettering equipment, and dry-mount press. Supplies include paper, ink, pencils, board, and shading film. Examples of general research expenses are publication subscriptions, books, and specimens.

No hourly rate should be determined without considering the fringe benefits that most artists in salaried positions receive. This includes paid vacation, sick leave and insurance programs, and prepaid trips to professional meetings.

The current salary range for illustrators at public institutions and the civil-service ratings for illustrators are a guide in computing your hourly rate. Be competitive without undercutting the established rate.

Expenses

Expenses that are applicable to a specific job are charged separately from those included in the hourly rate. These include any materials purchased especially for a particular job, such as phototypography, special papers, and films. It includes fees for models or expert advice, photography, duplicating, travel expenses, and telephone and postage costs.

Illustration inventory

For each project an illustration inventory should be kept. This is important for keeping track of the progress of the job, for making the estimate, and for retaining a permanent record of the job. Prepare a form similar to the following example and have it duplicated (12-3).

Contracts

An agreement to produce artwork for money constitutes a contract. It can consist of a telephone conversation, a discussion in person, a request in writing, or a comprehensive written contract. It is necessary that both parties agree on the extent of the arrangement. It is often advantageous for both artist and client for the responsibilities of each party to be spelled out in writing.

Points of agreement

Whether the contract is oral or written, there are certain points that should be covered:

1. Number of illustrations
2. Ultimate use of art (journal, book, slide, report)
3. Technique
4. Size (original and reproduction)
5. Lettering (cost, selection, pasteup; client's or artist's responsibility)
6. Reference and research material (furnished by client or artist)
7. Time schedule (conference time with client, client to check final sketches, due date for finished work)
8. Fees (should reflect your hourly rate plus expenses)

Form

While an oral agreement may often be sufficient, it is advantageous to send a summary of the agreement to the client after the preliminary conference so that all the salient points will be recalled correctly at a later time. This summary or memo can be extremely simple (12-4). Send this memo to the client immediately after the first conference, keeping a copy for yourself. A simple statement such as this can often prevent misunderstandings and avoid clouded business relations.

February 1, 1979

From: P. Wood

To: Dr. Gilbert

Re: Illustrations (five)

Pen and Ink

½ page each, Journal of Entomology

Preliminary drawings February 27

Finished drawings due March 5

Estimated cost: $325–$450

12-4.

CLIENT______________________________

USE OF ART______________________________

DATE IN________________ DATE DUE________________

CHAPTER FIGURE #	DESCRIPTION	PRINTED SIZE	TECHNIQUE	ESTIMATED HOURS	ESTIMATED EXPENSES	ACTUAL HOURS	ACTUAL EXPENSES	FINISHED

12-3.

A more inclusive artist-client agreement has been suggested by a group of members of the Association of Medical Illustrators: Neil O. Hardy, Robert J. Demarest, Carol Donner, and Gottfried Goldenberg. It covers many problem areas and may be used as a working basis for an agreement designed for your particular needs.

1. Name of client
2. Description of art
   - A. Subject
   - B. Number of units
   - C. Original size and reproduction size
   - D. Medium or technique
   - E. Method of reproduction
   - F. Specific use of art
     The artist and client should fully understand the use of the art. This should be in writing so that each party knows what constitutes a reuse.
   - G. Responsibility for placing type and leaders
     Artist
     Client
   - H. Reference material
     Supplied by artist
     Supplied by client
   - I. Artist and client agreement on signing of artwork or other mutually agreeable credit line.
3. Due dates
   - A. Sketches or storyboard
   - B. Finished art
     The due date should be established as an agreed-upon period of time after artist receives initialed and dated approved sketches.
4. Consultation time and/or observation of surgery
   "Off the board" time billed at hourly rate
5. Travel costs
   Expenses for any necessary travel should be billed separately or considered in the illustration price. This should be agreed upon before the project is initiated.
6. Corrections
   - A. No charge before sketch approval if changes are reasonable.
   - B. Sketches to be initialed and dated by client to indicate approval. Sketches should be returned promptly to artist. Due date for finished art should be based on receipt of approved sketches by artist.
   - C. Changes requested in finished drawings after sketch approval justify additional charge.
7. Price
   - A. Per-unit or hourly rate as agreed upon by artist and client. Any travel or per-diem charges should be included.
   - B. The artist reserves the right to renegotiate this contract after six (6) months of its submission. Present quotations are based on current salaries and operating costs.
8. Payment
   - A. Within thirty days of receipt of invoice for work satisfactorily completed, unless both parties agree to a different time period.
   - B. On large projects an advance payment should be considered.
9. Alterations
   - A. No alterations to artwork to be made without permission of the artist.
   - B. The artist's signature must not be removed from the artwork.
   - C. Any desired change should be made by the original artist or by an artist designated by him or her that is acceptable to the client.
10. Reuse
    - A. Use of the artwork beyond provision in Section 2-F constitutes a reuse. The artist to be reimbursed 25% of the purchase price for each and every reuse.
    - B. A mutually acceptable royalty arrangement is an alternative to the reuse provision.
11. Work stoppage
    It is understood and agreed that, once the work is started on the project by the artist, notice in writing is required to stop work. The client further agrees to be liable and to pay for all time and materials furnished by the artist to the project up to and including the day on which work is stopped in accordance with the written notice.
12. Termination
    This agreement may be terminated by either party upon written notice should the other party fail to perform in accordance with its terms through no fault of the other. In the event of termination the artist shall be compensated for time and materials expended up to and including date of receiving notice of termination.
13. Death
    It is understood and agreed that, in the event of the artist's death, his heirs and his estate shall be free of all liabilities for the incompleted portion of the contract. Work completed, if used, should be paid for.
14. Incapacitation
    It is understood and agreed that, in the event of serious illness or incapacitating injury to the artist, the artist reserves the right to terminate the contract or to appoint a duly authorized agent to complete the artist's portion of the subject contract. This authorized agent must be acceptable to the client.
15. Responsibility of artist
    All work sold as original must be the artist's creation and his copyrighted property. Avoiding plagiarism is the artist's responsibility.
16. Responsibility of client
    - A. Return artwork to artist in good condition upon completion of usage described in 2-D, unless all rights to artwork are sold. Artwork returned in damaged state shall be paid for at the same rate as it was originally.
    - B. Payment within stipulated period.

Bibliography

BOOKS:

1. GRAPHICS MASTER 2; Dean Phillip Lem, Dean Lem Associates, Los Angeles, California
2. NATURE DRAWING, Clare Walker Leslie; Prentice Hall, Inc., Englewood Cliffs, New Jersey, 1980
3. THE ARTIST IN THE SERVICE OF SCIENCE: edited by Walter Herdeg, The Graphis Press, Zurich, Switzerland, 1973
4. POCKET PAL, A GRAPHIC ARTS PRODUCTION HANDBOOK: International Paper Co., New York, N.Y.
5. DRAWING ON THE RIGHT SIDE OF THE BRAIN: Betty Edwards, J.P. Tarcher, Inc., Los Angeles, distributed by St. Martins Press, New York, 1979
6. THE ZEN OF SEEING; Frederick Franck, Vintage Books, a division of Random House, New York, 1973
7. SCIENTIFIC ILLUSTRATION: John L. Ridgeway, Stanford University Press, California, 1938 (reprinted 1979)
8. PASTE-UP; Rod van Uchelen, Van Nostrand Reinhold, 1976
9. PICTORIAL ANATOMY BOOKS (various); Stephen G. Gilbert, University of Washington Press, Seattle & London
10. MELLONI'S ILLUSTRATED MEDICAL DICTIONARY: Dox, Melloni, Eisner; Williams and Wilkins, Baltimore
11. NORTHWEST TREES; Arno & Hammerly, The Mountaineers, Seattle, Washington, 1977

MAGAZINES:

1. SCIENTIFIC AMERICAN, monthly, New York, N.Y.
2. COMMUNICATION ARTS MAGAZINE, bimonthly, Coyne & Blanchard, Inc. Palo Alto, California
3. GRAPHIS, bimonthly; Walter Herdeg, The Graphis Press, Zurich, Switzerland
4. NATIONAL GEOGRAPHIC, monthly, Washington D.C.

Index

acetate film *see* film
acrylic 60, 62, 67, 70, 71, 82
airbrush 51, 63–66, 67, 71, 82, 116
animals 78–83
atmospheric perspective *see* perspective, atmospheric

background 31, 32, 57, 61, 72, 82–83, 88, 91, 92, 93, 105
backpainting 61, 64, 71, 92
balance 96–97, 107
binding 116, 120–121, 132
black and white 20, 30, 32, 35–50, 67, 68, 82, 86, 87, 93, 104, 105, 114, 118, 126, 133, 136
blotting 60
board *see* bristol board, cardboard, coquille board, drawing board, foam-core board, hardboard, illustration board, pressed board, scratchboard, video board
books *see* print media
borders 96, 109, 114
boxes 96
brilliance 71, 72
bristol board 63, 77
brushes 37, 38, 43, 48, 49, 52, 53, 55, 56, 58, 59, 60, 62, 63, 64, 65, 66, 67, 70, 71, 137
budget 124, 127
business practices 143–145

camera lucida 23
captions 86, 96, 100, 102
carbon dust 51, 58–61, 64, 82, 116
cardboard 48, 110, 128, 130, 132, 136, 138
career guide 142–145
charts 93, 133
chroma 71
classification 78
clientele 143
collating 120
color 13, 14, 20, 21, 26, 28, 30, 35, 46, 51, 57, 61, 67–72, 78, 80, 82, 84, 86, 87, 88, 89, 91, 92, 93, 94, 95, 96, 104, 105, 107, 114, 118, 123, 126, 133, 134, 136, 137, 139
  complementary 71
  flat 67, 68–70, 86
  local 71, 72
  process 67, 68, 70, 114, 119, 132
  temperature 71
continuous tone 13, 30, 32, 44, 51–66, 68, 70, 82, 104, 105, 110, 112, 116, 118, 126
contour 20, 21, 28, 34, 38, 46, 60, 64, 80, 82
contracts 144–145
contrast 29, 58, 72
copy 70, 87, 118, 125, 138, 139, 140
copyright 122
coquille board 50, 82
core dark 27, 32, 56, 57
cover sheet 110, 114
crayons 50, 72
cropping 110, 114, 140
crosshatching 38, 42, 71
curves 88

depth 31, 57, 59
design 84, 96–115, 133–138
detail 14, 17, 18, 21, 28, 30–31, 34, 35, 39, 48, 51, 52, 57, 58, 59, 61, 63, 65, 71, 72, 80, 83, 86, 94, 96, 98, 106
diagrams 13, 84–95, 140
display *see* exhibition
distance 48
divider 23
  proportional 24
dots *see* stippling
Double Nothing 141
drafting film *see* film
drawing 12–25, 26, 31, 32, 38, 44, 45, 56, 58, 59, 60, 61–62, 65, 68, 70–72, 78, 83, 96, 106, 118, 125, 133, 139, 140
  board 14
  transfer *see* transferring
  tube 23
  wash *see* wash
dropout *see* halftone, outline
drybrush 57, 64, 72, 71, 82
dry mounting 48, 49, 89, 114, 141
dry transfer 91, 93, 104–105, 114, 136, 137, 138, 141
duotone 69, 70

electrophotography 118
engraving 35
enlargement 5, 17, 23, 24, 37, 92, 106, 107, 114, 116, 118, 136, 137, 138
erasers 14, 24, 37, 38, 48, 50, 52, 56, 57, 58, 59, 60, 62, 63, 65, 72, 87, 89
estimates 143–144
exhibition 67, 96, 118, 123–141
  assembly 128, 130, 132
  flat-hung 124, 127, 138
  packaging 128, 132, 139, 140
  self-standing 124, 128, 130
  storage 132
  tabletop 124, 128
  transport 128, 132, 139, 140
expenses 144
eye flow 96, 98, 107, 133, 134

film 24, 37, 49, 61, 62, 63, 64, 65, 66, 68, 87, 88–89, 91, 93, 94, 112, 114, 136, 140, 144
finishing 57, 60
fixative 58, 62, 105, 137
foam-core board 127, 128, 130, 136, 137, 138, 140
fold 119, 120–121
folio 106
font 104
foreground 31, 80
foreshortening 6, 8, 9, 10
form *see* shape
frisket 63, 64, 66, 70
frosted film *see* film
full color *see* color, process
function 14, 16, 78

galleys 122
glue 89, 110, 114, 137
gouache 72, 71
graphs 93–94, 139
gravure *see* intaglio
gray scale 30, 89
grid 24, 68, 96, 116
ground 37, 38, 50, 52, 58, 59, 60, 62, 63, 69, 71, 90
growth pattern *see* pattern, growth

halftones 50, 51, 110, 116, 118, 119
  outline 110, 112, 133
  square 110
hardboard 127
heading 127, 134
highlights 26, 29, 30, 31, 32, 56, 57, 58, 60, 65, 72, 77, 80, 110
hue 69,71, 72,

illustration board 37, 53, 56, 63, 64, 68, 70, 87, 91, 92, 114, 143
  cold-press 52, 71, 72, 77
  hot-press 52, 77
ink 37, 38, 39, 40, 43, 44, 45, 48, 49, 50, 57, 60,70, 72, 80, 82, 87, 88, 90, 91, 112, 114, 116, 118, 119, 120

intaglio 118
inventory 144

key line 68, 70, 110, 114
knives *see* scrapers

labels 69, 90, 93, 94, 96, 100, 101, 102, 105, 114, 125
layout 86, 87, 91, 96-115, 119, 123, 124, 136
lazy lucy 24
leaders 70, 100, 112
leading 104
lettering 86, 90, 91, 94, 100-103, 104-105, 114, 116, 124, 134-138, 141, 144
 hand- 105, 136-137
letterpress *see* relief
light 26-34, 48, 52, 56, 61, 82
 box 37, 91
 reflected 27, 56, 57, 65, 71
line 13, 31, 32, 37, 38, 39, 40-42, 43, 44, 45, 46, 59, 68, 70, 72, 82, 84, 87, 89, 104, 105, 106, 112, 116, 118, 119
 key *see* key line
 unweighted 42
 weighted 42
linear perspective *see* perspective, linear

maps 84, 94
margins 100, 102, 104, 106, 110, 112, 114, 120, 134
masks 59, 68, 69, 110
materials 37-38, 50, 52, 58, 62, 63-64, 87-90, 138-139, 140-141
measurement 21, 23-24
mechanicals 110-115, 116, 118
microscope 23, 24
modeling 13, 65, 70
mounting 62, 64, 65, 68, 71, 87, 91, 93, 110, 112, 114, 136, 138, 140-141, 142, 143
 dry *see* dry mounting
movies *see* projection media

negatives 70, 139

offset 118, 119
oils 72
outline 11, 13, 22, 33, 38, 39, 46, 50, 62, 70, 80
overlays 5, 22, 67, 68, 69, 70, 86, 104, 105, 109, 110, 112, 114, 118

paints 55, 57, 58, 60, 61, 62, 63, 64, 67, 70, 87, 91, 92, 136, 137, 138, 139
palette 52, 53, 64, 71
paper 5, 14, 35, 37, 38, 40, 48, 49, 50, 52, 59, 64, 66, 71, 72, 87, 89, 91, 92, 105, 107, 108, 116, 118, 119, 120, 132, 136, 138, 139, 141, 144
 cover 119, 120
 finish 119
 grain 119
 text 119
 weight 119
parallax 22, 80
parallelism 6, 8, 9, 10, 21, 32, 40, 48
pastels 72
pasteup 92, 93, 96, 104, 110, 114
pattern, growth 14-16, 21, 29, 49, 80
pencils 14, 24, 38, 44, 48, 49, 51, 52, 56, 57, 58, 60, 61-62, 63, 64, 67, 71, 72, 78, 82, 87, 91, 107, 114, 116, 118, 136
pens 37, 38, 40-42, 43, 46, 49, 50, 62, 65, 71, 87-88, 91, 93, 107, 114, 136, 137
perspective 6-11, 13, 21, 22, 80
photography 5, 12, 13, 14, 20, 24, 34, 51, 63, 68, 70, 78, 83, 89, 91, 92, 96, 101, 105, 107, 114-115, 124, 125, 126, 133, 140, 141, 142, 144
phototypography 104, 138
pica 102, 106
picture plane 6, 8, 9, 10, 21, 23
pigments 52, 53, 55, 56, 57, 63, 64, 71, 69
planography *see* offset
plastic 127, 128, 130, 136, 138-139
platemaking 5, 67, 69, 70, 77, 116, 118, 122
plywood 128
point 102, 104, 105
portfolio 142-143
position 17, 20
positives 91, 92-93, 136, 139
posterization 126, 132, 136, 137, 138
presentation 142-143
pressed board 127, 130, 138
printing 5, 69, 96, 110, 116-122 *see also* specific processes
print media 5, 67, 68, 70, 80, 82, 83, 84, 86, 89, 96, 98, 101, 102, 107, 126, 134, 144
projection media 5, 24, 67, 68, 70, 80, 82, 83, 84, 86, 89, 96, 98, 101, 102, 107, 126, 134, 144
proportion 86, 96, 106-107
 divider *see* divider, proportional
publications *see* print media

reduction 5, 17, 18, 23, 24, 35, 37, 51, 86, 92, 100, 102, 105, 106, 107, 114, 116, 118, 122, 136, 142
reflection *see* light, reflected
registration 68, 70, 110, 112, 114
relief 118, 119
reproduction 35, 51, 67-70, 144
reprography 118
resists 57
retouching 63
rhythm *see* pattern, growth
rubber cement 49, 57, 66, 110, 114, 137, 140, 141
rub-on symbols *see* dry transfer
scaling 86, 96, 106-107, 133
scrapers 37, 38, 48, 58, 60, 62, 63, 64, 65, 71, 87, 89, 90
scratchboard 37, 48-49, 71, 82
screens 5, 35, 51, 68, 70, 72, 110, 112, 114, 116, 118, 119, 126
 silk- *see* silkscreen
section 13
separation *see* color, process
shading 13, 18, 28, 32, 34, 37, 38, 44, 45, 46, 51, 56, 59, 61, 72, 82, 88-89, 93, 94
shadow 26-34, 48, 50, 52, 56, 57, 61, 65, 70, 80
 cast 18, 26, 31-32, 56, 57, 61, 69, 70
shape 14, 17, 20, 26, 27, 31, 32, 33, 38, 39, 46, 51, 52, 56-57, 59, 72, 80, 84, 96, 106, 127
signature 120
silhouette *see* halftone, outline
silkscreen 77, 118, 136, 138, 139
size 17-19, 20, 96, 98-99, 102, 106, 119, 127, 134, 144
slides *see* transparencies
space
 binocular 21
 monocular 21
 negative 13, 22
 positive 13, 22
spacing 102, 104
specimen *see* subject
stippling 37, 38, 39, 46-47, 50, 82, 89
stomping 60, 61
structure 14, 16, 38, 67, 72, 78, 80
studio 14
subject 13, 14-16, 17, 18, 20-24, 26, 27, 28, 31, 33, 34, 38, 39, 48, 52, 55, 57, 59, 61, 71, 72, 78, 83, 96, 98, 142, 144
surface texture *see* texture

tables 84, 133
tapes 89-90, 91, 141
television *see* projection media
templates 87, 88, 91, 114, 137
text *see* copy
texture 14, 20, 21, 28, 29, 38, 46, 51, 59, 61, 65, 69, 71
three-dimensionality 13, 21, 30, 31, 34, 48, 133
titles 125, 127, 134
tone *see* continuous tone
tortillon 58, 60, 61, 62, 77
tracing 22, 24, 35, 37, 40, 44, 49, 56, 62, 64, 66, 70, 87, 91, 92, 107, 108, 136, 137
transferring 24-25, 35, 49, 50, 56, 59, 61, 62, 64, 70, 91, 92
 double 24, 37, 48, 56, 62
 dry *see* dry transfer
 single 24, 37, 56, 62, 72
translucence 32, 87
transparencies 5, 67, 70, 83, 84, 86, 87, 91, 96, 105, 106, 107, 126, 134, 139, 142, 144
tripletone 70
typefaces 102, 104, 134
typography 5, 124

value 11, 24, 27-31, 35, 37, 46, 50, 51, 52, 55, 56, 57, 59, 60, 61, 62, 63, 64, 65, 66, 71, 70, 80, 114
vanishing point 8, 9, 22
vellum 24, 37
video board 37, 71
videotape *see* projection media

wash 37, 51, 52-57, 71, 82, 116
 graded 55-56, 57, 63, 71, 72
 smooth 53-55, 56, 57, 63, 71, 72
watercolor 51, 52, 63, 64, 65, 67, 69-71, 78, 82, 87
 wet-on-wet 55
wax-back symbols *see* dry transfer